VITAMINS AND MINERALS IN HEALTH AND NUTRITION

MATTI TOLONEN

ELLIS HORWOOD
NEW YORK   LONDON   TORONTO   SYDNEY   TOKYO   SINGAPORE

First published in 1990 by
ELLIS HORWOOD LIMITED
Market Cross House, Cooper Street,
Chichester, West Sussex, PO19 1EB, England

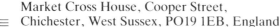

A division of
Simon & Schuster International Group

© Ellis Horwood Limited, 1990

Typeset in Times by Ellis Horwood Limited
Printed and bound in Great Britain
by Hartnolls Limited, Bodmin, Cornwall

Exclusive distribution by Van Nostrand Reinhold/AVI London:

Australia and New Zealand:
CHAPMAN AND HALL AUSTRALIA
Box 4725, 480 La Trobe Steet, Melbourne 3001, Victoria, Australia

Canada:
NELSON CANADA
1120 Birchmount Road, Scarborough, Ontario, Canada, M1K 5G4

Europe, Middle East and Africa:
VAN NOSTRAND REINHOLD/AVI LONDON
11 New Fetter Lane, London EC4P 4EE, England

North America:
VAN NOSTRAND REINHOLD/AVI NEW YORK
115 Fifth Avenue, 4th Floor, New York, New York 10003, USA

Rest of the world:
THOMSON INTERNATIONAL PUBLISHING
10 Davis Drive, Belmont, California 94002, USA

British Library Cataloguing in Publication Data

Tolonen, Matti
Vitamins and minerals in health and nutrition.
1. Man. Nutrients: Minerals and vitamins
I. Title
613.28
ISBN 0–7476–0068–6

Library of Congress Cataloging-in-Publication Data

Tolonen, Matti.
Vitamins and minerals in health and nutrition / Matti Tolonen.
p. cm. — (Ellis Horwood series in food science and technology)
ISBN 0–7476–0068–6
1. Nutrition. 2. Vitamins — Health aspects. 3. Minerals in the body.
I. Title. II. Series.
[DNLM: 1. Health. 2. Minerals. 3. Nutrition. 4. Vitamins. QU160 T653v]
QP141.T73   1990
612.3′92–dc20
DLC
for Library of Congress
90–4617
CIP

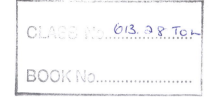

VITAMINS AND MINERALS IN
HEALTH AND NUTRITION

Table of contents

Preface

I don't want to convince. I want to stimulate new ideas and break down prejudices.
Sigmund Freud (1856–1939)

In the course of the past few years, research on nutritional medicine has proceeded dramatically; today there are virtually no researchers — neither doctors nor nutritionists — who would dare to claim that he or she has read, or has a complete command of the whole area concerning nutritional medicine, encompassing vitamins, minerals, amino acids and essential fatty acids.

This development in nutritional medicine has taken place in a way which is analogous to the process of the peroxidation of lipids: first there is a very long latent period where almost nothing happens. Then suddenly lipid peroxidation starts — the process proceeds with remarkable rapidity and nothing can stop it. Research into the role of free radicals and antioxidants on health can be described in the same terms. We find ourselves, at the present moment in time, at the beginning of a rapid chain reaction, which is spreading within the cell (e.g. in a country like Britain) and from cell to cell (from country to country). More conferences on free radicals, vitamins, trace minerals, prostaglandins and essential fatty acids are being held than any one individual can ever hope to participate in. Nobody has time to read all the literature on these subjects, so crucial to our health. Researchers often find themselves leafing through the collections of abstracts and reviews instead of the original publications — and they are always a few steps behind the absolute front line of research.

I am well aware of the fact that this book cannot possibly be a complete presentation of everything which has been published on nutritional medicine, vitamins, minerals, amino acids and essential fatty acids. In fact, I have attempted to edit the book wherever possible. I have included a range of new results, which have been collected variously from journals, international conferences, lectures and conversations with eminent colleagues. I have not even attempted to document all the information with references to the literature, otherwise the book would have been cumbersome to read. Those readers (researchers in particular), who wish to dig deeper into specific topics, have today access to databases with updated literature sources such as Medlars, Medline, etc.

Above all, I hope that this book, of which 100 000 copies have already been printed in Scandinavia, will be of use to everyone who wants to look after their health, or take care of their ailments, with vitamins, minerals, essential fatty acids and, in general, a healthy, balanced diet.

Espoo,
October 1989

M. Tolonen

Introduction

The simple and the dead never change their opinion.
James Russel Lowell (1819–1891)

Research in the field of nutritional medicine has opened up new perspectives for the prevention and adjunct treatment of a range of common diseases and of premature ageing. It is now becoming increasingly evident that there are a number of micronutrients which are of immense importance for our health — not only for the prevention of classical deficiency disorders like scurvy but also for prophylaxis of cancer, cardiovascular diseases, autoimmune diseases, cataracts and many other diseases. Similarly, it would appear that micronutrient requirements are greater for those who suffer from chronic diseases and who are under medication. The reason for this is that most diseases develop gradually by producing dangerous rancid lipids, lipid peroxides, which in turn give rise to further cell damage. This is where cell-protective vitamins, minerals, amino and fatty acids come into the picture: they are able to break the vicious circle.

Recently, it has become much more common for medical journals to contain references to the importance of diet in health and disease. The quantity of published literature relevant to nutritional medicine is extensive, probably in the region of 8000 to 10000 articles on an annual basis. Because of their inadequate training in nutritional medicine and because of the natural delay in the clinical application of the new knowledge to individual patients, most doctors have difficulty in answering nutritional questions correctly. They often know no more than their patients — and sometimes even less.

The doctor can find him- or herself in an embarrassing situation when a patient asks for nutritional guidance, and he or she has no idea how to answer. The situation is exacerbated because the patient expects the doctor to be an expert in the field. On the other hand, it is difficult for the doctor to admit to a lack of nutritional knowledge, because he is supposed to represent all the scientific knowledge available on health, including nutritional aspects, and because he is afraid to lose the patient's confidence. This dilemma can result in the doctor attempting to save face by claiming, for example, that selenium and other trace elements are just humbug; or he might say that everybody receives sufficient minerals and vitamins from their diets. Nutritional guidance can therefore end up with the patient simply being told to eat whatever he likes best. This hardly leaves the patient any the wiser.

Many find it difficult to understand that doctors and nutritionists can disagree to such an extent about the effects of vitamins and trace elements. Traditionally, most doctors are of the opinion that dietary supplementation is quite unnecessary. Others, however, diagnose trace mineral deficiency and recommend supplementation. How are we to know who is right? Is what was true yesterday no longer true today? Do the food researchers know more about the importance of diet, in questions of health and illness, than doctors do? Are nutritional questions the domain of doctors or of nutritional science — or of both? Who are the real experts?

The need for dietary supplements of micronutrients is a matter of heated debate all over the world. The difficulty in the debate about dietary supplements is that the psyche of the debater is intermingled with the empirical truth under argumentation. It would seem that the time is now ripe to abandon some of the traditional ways of thought which have hitherto been regarded as unimpeachable. Orthodox medicine and food science traditionally held the view that we must get all the elements from food and therefore food supplements have been denied any value. It is also feared that the use of food supplements would lead to food faddism and negligence of well-balanced diets. At the same time more and more studies reveal overt and latent micronutrient deficiencies in large portions of populations, clearly indicating that the slogan 'we get everything we need from food' is simply obsolete.

In many countries the food industry fortifies some foodstuffs, e.g. milk, corn flakes and flour, with vitamins and iron. Now the question arises, what is the difference between supplementing foodstuffs and taking pills? From the biochemical and functional point of view, it does not matter whether the cells get their iron and vitamins from food or pills; the only thing that matters is that the cell gets enough of the essential element.

From the point of view of the doctor and the patient, dietary supplements in the form of pills or capsules are the quickest and most efficient way to correct a deficiency. This is how iron or potassium deficiency is treated in the best medical practice. No doctor would rely on dietary counselling in treatment of overt iron deficiency. Alas, selenium or vitamins are, illogically enough, banned by orthodox medicine.

It is not my intention to belittle the significance of nutritional guidance — quite the contrary, as you will see throughout this book. But I wanted to point out the emotional, non-intellectual attitude often exhibited in articles and discussions dealing with dietary supplements.

Nutritional medicine

In 1984 the British Society for Nutritional Medicine (BSNM) was established in the UK to promote awareness about nutritional medicine amongst members of the medical profession and the lay public.

According to BSMN, **nutrition** is defined as 'the sum of the processes involved in taking in nutrients, assimilating and utilizing them'.

Nutritional medicine is defined by BSNM as 'the study of interactions of nutritional factors with human biochemistry, physiology and anatomy, and how the clinical application of a knowledge of these interactions can be used in the modulation of structure and function for the prevention and treatment of disease as well as the betterment of health'.

Prime principles of nutritional medicine include, according to BSNM, the following factors.

(1) Man's diet, even in industrialized societies, may very often have only a borderline or indeed low content of certain essential nutrients. A 'normal' diet is not necessarily a healthy or optimum one.
(2) Requirements for essential nutrients vary from individual to individual depending on genetic, physiological, lifestyle and other influences. What is adequate for one person may not be for another.
(3) Illness is inevitably linked with an abnormal biochemistry and an alteration in the metabolism of nutrients and their byproducts.
(4) Specific nutrients such as vitamins, minerals, essential fatty acids and amino acids, as well as dietary manipulation in general, provide a potent means of influencing body chemistry and thus disease processes.
(5) By correcting fundamental biochemical abnormalities by nutritional means one can prevent certain diseases or alter the course of a disease for the better.

Dr **Stephen Davies**, M.A., B.M., B.Ch., Chairman of BSNM, stated recently at a medical conference in Helsinki:

> While all doctors have made use of and continue to rely on drug-based therapies, there are many clinical situations now known to be amenable to a nutritional approach, with resultant improvement in patient care and a reduction in drug reliance.
>
> In western society the remarkable general neglect within clinical medicine of the importance of nutrition is probably based on the doctors' failure to comprehend important fundamental concepts of nutritional medicine.

Nutritional medicine is still in its infancy and researchers are still divergent in their opinions; nevertheless it is important that the individual gathers the necessary available information in order to affect his own health and well-being, and that of his friends and family, in the right direction. This demands that we are informed about the latest research and practical experience.

Note to the reader
In order to make for easier reading, I have consistently used the word mineral to describe the whole group of minerals and trace elements.

In the text, the words *serum*, *plasma*, and *whole blood* crop up on a number of occasions. *Serum* is that part of the blood which does not contain blood corpuscles. In contrast to blood *plasma*, serum contains no fibrinogens (coagulating factors). *Whole blood* is blood from which none of the component parts have been removed.

There are a number of cross-references within the book; these are mostly standardized: for example, (see 2.6) means look up Chapter 2, section 6.

A glossary of medical terms can be found on page 205.

1

Take a new look at your life

1. HOW LIVING HABITS AFFECT OUR HEALTH

The best doctors in the world are Doctor Diet, Doctor Quiet and Doctor Merryman.

Jonathan Swift (1667–1745)

Two decisive factors have an influence on human health: on the one hand, the inherited characteristics (genes), and on the other, living habits and environment. We can do nothing about our genetic inheritance. On the other hand, there is a great deal we can do for our health through our choice of lifestyle and living habits. We can reduce the risk of illness or we can affect the course of an illness which we already have. Besides, we would all like to age slowly and with dignity.

It is here that diet and nutrition come into the picture — and with them come calories, vitamins, minerals, amino acids and essential fatty acids. Diet has far more to do with illness and the ageing process itself than was previously believed. Doctors usually only warn against smoking, excessive intake of alcohol and a few other damaging habits. Only now have doctors begun to see a connection between diet and illness.

Our genes determine our physical and psychological structure. We can inherit diseases, but we can also inherit a tendency to contract a specific disease if a range of other environmental factors are present. This is the case with allergies, for example, which can break out if there are allergens present in the environment or in food.

The genes also decide how long our biological clock can function, i.e. they determine the length of our maximum life span. It would appear that a great deal is predetermined; but all is not lost — everyone holds the key to their own happiness, at least partly. This is especially true of the field of health, where the individual has a wide range of possibilities for taking action. It is possible to choose healthy or unhealthy living habits. Our health is influenced by a vast number of factors: the quality of food, eating habits, coffee, tobacco, alcohol, work, exercise and sleep, to name only a few examples.

Psychological stress is the source of many complaints. Here the genes probably play a role too: we inherit the ability to tolerate a certain amount of stress. With this follows a tendency either to seek or to avoid high-stress situations. Many people become stressed simply because they take on too much activity. They cannot bring themselves to say 'No'. If the individual in question is one of those who have a high tolerance for stress then it is of no consequence. If not, then the individual is likely to exhibit the so-called psychosomatic symptoms: migraine, dizziness, muscle-tension in the head and neck, palpitations of the heart, pressure in the chest, breathing difficulties, undefined stomach pains and the like.

In 1935, **Hans Selye** (1907–1982), an Austrian biochemist, was the first to use the word *stress* as a medical term. Prior to this, stress was used to describe technical strain. According to the latest research, stress not only affects the hormones (adrenaline, etc.) and the nerves, but also the immune system (which is delineated in section 6 of this chapter). Stress is usually described in pejorative terms, but this is not necessarily the whole truth. Recent studies show that an optimal stress increases the production of antibodies in the body, while a too strong and long-standing stress may decrease antibodies.

American professor, **Chandler Brooks**, one of the pioneers of physiology, presumes that this may partly explain why widows often die shortly after their husbands pass away. Dr Brooks expressed these views at the 100th Anniversary meeting of physiology, in 1989 in Helsinki.

Stress, experienced as mental overload, has increased tremendously. Strenuous stress is often accompanied by other unhealthy habits. You don't have time to eat regularly, you drink too much coffee, perhaps you drink too much alcohol and smoke too much. The punishment usually comes the next day: you are tired and sluggish and you try to forestall this by taking more stimulants. If you have landed in this sort of vicious circle, then it is high time to make a concerted effort to get out of it again. It is just this kind of lifestyle which sooner or later inevitably leads to vitamin and mineral deficiency — either because you don't get enough through your diet, or because the coffee, cigarettes and alcohol steal the essential vitamins, minerals and trace elements.

In our modern industrial world, anaemia (iron deficiency) is common — it arises quite simply out of an unbalanced diet. At first sight, this may seem unbelievable to the reader but it is a well-established fact. In Sweden, known for its high standard of living, 7% of women and 5% of teenage girls suffer from overt anaemia. Iron deficiency, seen in blood tests, occurs in 30% of Swedish women and 35% of teenage girls. Overt anaemia has been reported in 6.4% of Finnish women, 1.3% of French women and 2.4% of German women. Anaemia can develop in women of all ages and in middle-aged men — the individuals in question lack either the time or the desire to ensure that their diet is sufficiently varied. Elderly people often eat very little fish, meat or vegetables, sometimes to save money, sometimes because their teeth are not as strong as they once were. This leads to an insufficient daily intake of vitamins and minerals — contrary to what was previously believed.

As a doctor, I see many overweight people who are convinced that they are overweight because of a problem in their hormonal metabolism. 'I eat like a sparrow,' they often say. So we undertake all sorts of investigations and laboratory tests which reveal that there is, in fact, no metabolic disorder whatsoever. One becomes overweight quite simply because one eats more than one can use up. It would appear that many people eat a large amount of food without noticing it, while nobody quite unconsciously takes a lot of exercise. Obesity is a contributing factor (risk factor) in a great number of illnesses: high blood pressure, heart conditions, diabetes and arthritis in the hip or knee joints, to name but a few.

We are often encouraged to go for regular medical check-ups, much in the same way as we send the car in for regular servicing. The analogy is perhaps not the best. It would be more appropriate to compare it with an M.O.T. Here, if there is something wrong with the car, it is the owner's, not the tester's, responsibility to ensure that problems are attended to. Similarly, it is our own responsibility to ensure that we do something about the problems and deficiencies which the doctor unearths at a medical. But how closely do you follow the doctor's advice? That is, if you go to regular medical check-ups at all.

Doctors are supposed to know more about health than laymen. As a matter of fact, it would indeed appear that doctors also enjoy better general health. The standardized death rates (cases, per 100 000 persons in a certain age group) for doctors are indeed much lower than the average for the population at large. What is the explanation for the doctors' better health? The single most significant factor is presumably that doctors smoke less. Furthermore, it has been ascertained that doctors have lower blood pressure and blood cholesterol levels than the general population. This must surely be due to better nutrition and healthier living habits in general.

It is worth noting that young doctors are more frequently in the forefront than their elder colleagues, when it comes to the promotion of changes in living habits and enhancing the national health through preventive measures.

2. SELF-TREATMENT — ADVANTAGES AND DANGERS

Sickness is felt, but health not at all

Thomas Fuller (1608–1661)

We deal with most of our ailments and illnesses ourselves. Mr or Ms Average only visits the doctor two to three times in a year. On the other hand, we make much more frequent use of the medicine cabinet or the chemist. Health food shops are also becoming increasingly popular. The modern man or woman often prefers natural methods of treatment, as opposed to synthetic medicaments. But we cannot escape the fact that medicine is an important part of our lives. Self-treatment has many advantages, but there are also a number of pitfalls associated with it.

The concept of self-treatment stems from the 1970s. It implies quite straightforwardly, that the individual attempts to treat his illness himself, without visiting the doctor. Ninety per cent of all ailments are today treated by the patient him- or herself.

OTC (over the counter) medicines which we buy from chemist's shops are used for self-treatment. It is still a rarity for a doctor to give a patient advice on specific products from the available selection in the health food shops. With the passage of time, vitamins and minerals have come to represent a considerable proportion of those products used in self-treatment, particularly in the USA.

Self-treatment has a number of advantages to recommend it — not least of an economic nature. The entire budget of the National Health Service would have to be more than doubled if most of us consulted the doctor at the slightest symptom of illness. Even greater savings could be made if self-treatment was rationalized — it could be a much more effective means of combatting illness if more people were given access to the necessary and relevant information, which would enable them to deal with even more ailments without having to visit the doctor. With the correct information at his fingertips, the patient need take no risks in his treatment.

There are, however, a number of obvious dangers associated with self-treatment. One could envisage, for example, a situation where the patient treats his illness himself for too long before going to the doctor. In such a case, it is possible for a serious illness to become so advanced, that nothing more can be done about it. It ought, therefore, to be the accepted practice, that the doctor makes a diagnosis which can serve as a guideline for the most effective course of treatment for the patient. To ignore the doctor's advice can have disastrous consequences in cases of serious illness, cancer for example. The patient should, in the first place, never put his trust in naturopaths, wise old ladies or other 'good advice'.

On the other hand, the personnel at the chemist's can assist the patient in the appropriate choice of treatment in many situations; the same applies for many health food shops, although here the quality of advice can vary more widely. This problem has, however, been alleviated by educational improvements in recent years.

Some patients, unfortunately, still have the impression that if the doctor has prescribed a certain dosage of a medicine, then a self-administered double dosage of the same must be twice as beneficial. Take care! There is no foodstuff, food supplement or medicine which cannot be dangerous when consumed in massively exaggerated quantities. Excessive consumption of vitamins, minerals and essential fatty acids can also do more harm than good, because the body can only make use of a certain amount of them.

We have hitherto been used to the assumption that the water-soluble B vitamins are harmless because they are eliminated through the urine within 6–8 hours after intake. However, cases of serious vitamin B poisoning have been reported in the United States following enormous overdoses of up to 1000 times the Recommended Dietary Allowance (RDA) (see 1.7) of, for example, vitamin B6. Vitamin A poisoning has also been detected, but this has arisen in cases where individuals have ingested from 30 000, to up to more than 100 000 international units (IU). The recommended daily dosage for vitamin A is 5000 IU. It is characteristic of vitamin A poisoning that extreme overdosing has continued over an extended period of months, or even years. Pregnant women should furthermore be especially cautious with regard to overdoses of vitamin A, owing to the possible risks of embryo damage. The maximum safe daily intake of vitamin A during pregnancy is considered to be 8000 IU. During pregnancy, β-carotene a precursor to vitamin A, which is found in carrots — and which is neither teratogenic nor toxic — is to be preferred.

Vitamin E, when taken orally, has never been the cause of actual poisoning in human beings, although doses of over 400 IU per day can give rise to stomach trouble. However, when such high doses are used, it is always best to seek the advice of a doctor who is well-informed in the field of vitamins. This ought also to be the rule in the case of other, more planned, courses of treatment; unfortunately, there are still far too few doctors who have a thorough command of this field. Nevertheless, if one follows the recommended dosages on the labels or bottles of the products from the chemist or the health food shop, there will be no problems. When a doctor, after careful consideration, prescribes a megadose of a medicine (or of vitamins or minerals), one ought to be able to count on, that he will supervise the treatment, and that he will intervene at any sign of side-effects.

Vitamin C is neither harmful nor toxic, even in very large doses. But because ascorbic acid is an acid, it can lead to stomach irritation, or, at the worst, stomach ulcers, if one has a tendency for them.

A rule of thumb for all vitamins is that a ten-fold dosage of fat-soluble, and a fifty-fold dosage of water-soluble vitamins can result in poisoning.

Selenium is the most talked-about trace element these days: one can take up to 200 micrograms of organic selenium per day, in the form of a food supplement, without any risk. This amount in no way exceeds the natural selenium intake in many areas in Japan, USA, Canada and South America. In the state of South Dakota the average daily intake is 400 µg, in Costa Rica it is all of 600 µg and in certain other parts of the world it is even higher — without any signs of health risks having been established.

In China there are areas with both very high, and very low selenium intakes: in Enshi-Hubei province selenium poisoning has occurred, but this was at a daily intake of 900–5000 µg.

If one takes a daily food-supplement in the form of polyunsaturated fatty acids, such as EPA, DHA or gamma-linolenic acid (γ-linolenic acid, GLA), then it is extremely important that one protects the fats against oxidation with antioxidants, e.g. selenium, zinc, and the vitamins, A, B6, C and E. Without the protection of antioxidants, the blood content of the so-called lipid peroxides (waste products from the oxidation of fats) increases and this can be dangerous.

3. WHY HAS NUTRITIONAL MEDICINE BECOME SO POPULAR?

Let nature work freely in everything

Jean-Jacques Rousseau (1712–1778)

The advantages and disadvantages of what is called biological or orthomolecular medicine have recently been the objects of a lively debate in most European countries. However, in the heat of the debate for and against, it is often forgotten that 95% of the current therapies of the best medical practice have not been tested in double-blind studies, and that normal medicines were originally produced from naturally occurring substances and herbs. Environmental protection, organic farming, health foods and the treatment of illness in accordance with biological principles are all concepts we have come to be familiar with. They are a reflection of our growing anxieties about pollution, herbicides, food additives and the increasing use of chemicals in the environment. There is also a widespread awareness of the side-effects of medicines.

A few years ago, I was employed by the World Health Organization in Geneva and often met a family from Finland. I was astonished when I heard that a Swiss doctor treated the daughter with magnesium for undefined fatigue. At the time I thought that this sounded quite mad. Fortunately, I kept my thoughts about my Swiss colleague to myself: the fatigue was in fact cured, and the treatment had been both cheap and risk-free.

Without doubt a wide range of herbal medicines also have medicinal effects. A quarter of all prescribed medicines derive originally from herbal or plant preparations: digitalis (digoxin), ephedrine, codeine, atropine, morphine and the migraine medicine ergotamine are but a few examples. If we include all the synthetic medicines which formerly were made from plants and herbs, then we can ascertain that more than half of the assortment on the chemist's shelves has its derivation in nature.

There are very few doctors who think about this, either because they lack sufficient knowledge of natural medicine, or because they are no longer aware of the origins of the products they prescribe.

The opponents of micronutrient therapies regard them as being ineffectual, or they maintain that positive results are merely products of what is called a placebo effect (see 1.9). Research lags so far behind the results, making it difficult to disprove these claims. In many countries, however, research is beginning to get under way into the effects of micronutrient products — or the lack of effects. There is a rapidly growing interest in the products at medical educational institutions; many young doctors are starting research in this field immediately after their graduation.

Considering the negative attitude of most doctors to nutritional medicine and dietary supplements, it is hardly surprising that many patients find it wiser to keep it to themselves that they use the likes of vitamins and minerals. Some doctors go so far as to make fun of the patients' attempts to cure themselves by natural methods. Fortunately for the patient, it is of no consequence to what degree the treatment is 'scientific', or tried and tested in double-blind trials, as long as he or she is cured. It would indeed be a healthy development, if the patient was able to discuss these questions openly with the doctor. The patient would thereby develop more confidence in the doctor, and this would reduce the likelihood of the patient giving up vital medical treatment because of side-effects.

Scientific testing and the placebo effect will be discussed in more detail in section 9 of this chapter.

4.   THE LIFE AND DEATH OF THE CELLS

Knowledge is food for the soul

Plato (428–347 A.D.)

The biological clock
All of our cells are programmed from birth to live for a specific period of time. Some cells are only intended to function for a relatively short time. Those cells which are concerned with the vital functions, like the brain cells, should, however, preferably remain healthy well into old age. It is possible for these cells to live for about 110–120

years, but not everybody grows to such a ripe old age; our way of life exposes us to a great number of harmful substances which damage the cells. This in turn leads to illnesses which shorten our life-span.

Many men grow bald, and womens' production of sex hormones ceases with the menopause. These are examples of cells which are designed to function for only a part of our life-span. Genetic factors determine what proportion of a lifetime they continue to function.

The older a woman is when she gives birth, the greater the risk is that she will have a child with congenital defects. These foetal defects are caused by the genetic material in the egg cells, which consist of DNA molecules. When a woman reaches the age of about 40 years, the cells gradually begin to lose the capacity for reproduction; it is in this period that significant changes occur in the egg cells and the risk that defects will be transferred to the foetus increases.

A human being's cells live for as long as the biological clock in them is programmed to keep ticking — if we ensure that we keep them healthy. When the machinery seizes up, the cells die, and the person dies with them. This is, in any case, nature's intention. Many scientists and researchers have sought for a means to slow the clock down, or even to make it go in reverse. This search has so far failed to produce any definite results, but there are indications which suggest that it may be possible to slow down a pathological ageing process by affecting the factors which cause the biological clock to tick faster. Amongst these damaging factors are the oxygen free radicals, lipid peroxidation, degradation of the protein molecules and cell changes, called mutations. We shall examine in more detail below, these processes, whereby the cells undergo damage and death. I describe these changes as fortuitous, in contradistinction to cell changes which are determined by nature. It is precisely these fortuitous mutations which give rise to illness.

Oxygen free radicals

Free radicals are chemical species with one or more unpaired electrons in their outer orbital. Our body normally produces them, since they are essential to normal bodily functions, but an uncontrolled generation of free radicals is destructive to the cells, tissues and organs. We review the free radical production due to external and internal sources and the intra- and extracellular defence systems which protect the cells against free radical toxicity. Today, most diseases are linked with free radicals and their antidotes, the antioxidants.

Oxygen is, at one and the same time, both essential to life and poisonous. At first sight this statement may sound strange. If so, this is because you have not heard of the dangerous free radicals of oxygen. We are all familiar with the life-supporting aspects of oxygen. Our cells need the oxygen contained in air to produce energy; the more efficient the absorption of oxygen, the better our fitness. Through regular exercise we are able to increase our oxygen absorption and improve our aerobic capacity. Oxygen, however, has properties other than energy production. Oxygen free radicals are oxygen atoms which have an unpaired electron. This leads to the formation of compounds which are harmful to health. The most commonly occurring oxygen free radicals are:

— superoxide
— hydrogen peroxide
— hydroxyl radical
— singlet oxygen.

Each of the free radicals of oxygen exists for an infinitesimally short period of time, perhaps only a fraction of a second, but this is long enough for them to attack the cellular structure and cause serious damage. The cell membranes consist largely of fats, more precisely of fatty acids. We are all aware of how fats can go rancid: this should make it easier for us to understand how the cell membranes are especially sensitive to the rancidifying effects of the free radicals of oxygen. Researchers are now in agreement about the deleterious effects of the free radicals of oxygen. They are the commonest cause of fortuitous mutations, which are associated with more than 60 different diseases, ageing itself and death. The free radicals not only cause lipid peroxidation (rancidification of cell membranes) but they also precipitate the following problems:

● cross linkages in the connective tissue, i.e. collagen and elastin. Around 30% of the protein in the body is comprised of collagen, which is located in the bones, the muscles, the sinews and the cartilage. Elastin is also a protein, and is found in, amongst other places, the artery walls. If cross linkages occur in these proteins, the skin and other connective tissues become stiff and age rapidly. Cross linkages also lead to changes in the blood vessels and to hardening of the arteries (arteriosclerosis).
● oxidation and destruction of large carbohydrate molecules from which mucous is formed. Mucous acts as a lubricant for the joints.
● increased quantities of the ageing pigment (melamine, lipofuscin and ceroid) in the skin, in the internal organs, in the nerves and in the grey matter of the brain.
● damage to the cell membranes due to rancidification of fats. Free radicals of oxygen form peroxides in the cell membranes, causing their destruction and leading to the formation of new oxygen radicals. This chain reaction eventually leads to the destruction of the cell.
● peroxidation products, e.g. oxidized LDL-cholesterol, circulate with the blood, giving rise to circulatory disorders. These substances also irritate the walls of the larger arteries and inhibit the formation of prostacyclin (PGI2), a beneficial substance which itself inhibits the formation of blood clots.
● rancidified fats within the cell are broken down and form harmful prostaglandins and malonaldehyde, a toxic compound which can effect mutations and cross linkages.

The free radicals are like an inner radiation, to which our cells are exposed ten thousand times a day throughout life. If our cells are not well-protected by antioxidants to impede this radiation, then the cells become sick and ultimately die.

An extreme example of an illness which exhibits the characteristics of the above-mentioned changes is the extremely rare disease *progeria*. Progeria causes the cells of young children to age extremely rapidly, usually resulting in the child's death from 'old age' before puberty. Another, more common, illness in children is *juvenile neuronal lipofuscinosis*, which is caused by an accumulation of age pigment in the nerve cells. This disease has been treated successfully in Finland with supplements of selenium, vitamin E and vitamin B6, called Westermarck's formula.

Overweight people have greater quantities of rancid fats in the body than thin people. This is why they also run a greater risk of contracting arteriosclerosis and heart infarction. The risk of contracting cancer increases as we grow older — the risk is directly proportional to the number of years raised to the power four. In experiments with laboratory animals (mice, for example), it has been possible to increase their life-span and prevent cancer by caloric restriction and by feeding them antioxidants, which counteract lipid peroxidation (rancidification). Restriction of calories also increases the life-span of animals.

The natural dietary antioxidants selenium, zinc, vitamins A, C and E plus β-carotene protect against free radicals, lipid peroxidation and even against *cancer*, as will become clear from the next chapter. For the moment, however, I want to deal with some examples of other diseases which are precipitated by the activity of free radicals.

Excess rancidified fats (lipid peroxides) are found in the tissue surrounding an area of cell damage, after a *coronary thrombosis*, for example. Free radicals damage heart cells in exactly the same way as they do any other cells which are diseased. When the coronary arteries are constricted, it gives rise to oxygen deficiency; when the blood flow is again normalized, the formation of free radicals of oxygen increases significantly. This phenomenon is called *reperfusion injury* (perfusion means blood flow) and damages the heart-muscle cells, thus increasing the risk of another heart attack or cardiac arrhythmia. Experiments with laboratory animals, in which synthetic or natural antioxidants have been administered, have been successful in protecting the heart muscle.

Brain tissue contains a lot of unsaturated fatty acids (lecithin, for example) which may become rancid. Rancidification, in turn, results in increased quantities of lipofuscin (age pigment) in the brain; in large quantities this age pigment speeds up the ageing process with the result that the patient becomes prematurely senile. In *Alzheimer's disease*, a certain group of brain cells become dark brown as a result of oxidized fat.

It would appear that the common eye disease, *cataract*, is also caused by a flaw in the oxygen metabolism due to the effects of light and free radicals. Experimental and epidemiological evidence indicates that antioxidants, i.e. selenium, β-carotene, vitamin C and vitamin E, protect us from cataract. Also *retinal cells* are very vulnerable in the event of (especially) selenium and vitamin E deficiencies. This is the reason why antioxidants accumulate in large quantities in the eye, in order to protect against possible oxygen-derived free radical damage.

Free radicals can also arise in connection with oxygen overdosing in an incubator. In such cases free radicals have led to eye problems for the premature babies. It is, however, possible to protect the baby's eyesight with vitamin E supplementation, which counteracts the oxygen radicals.

There is, by now, a massive body of scientific evidence to show that *most chronic and long-term illnesses* are associated with free radicals.

Now that we have a clearer picture of how diseases originate, it is possible to treat them more effectively — and perhaps even to prevent them from arising in the first place.

The harmful effects of the free radicals were discovered in the 1930s in the oil, rubber and plastics industries, but doctors and biochemists have only begun to take an interest in them in the 1980s. It may seem surprising that it has taken so long for this vital information to reach us; however, few members of the medical profession are aware of the significance of free radicals for the origins and exacerbation of illness.

Dr **Denham Harman**, who in 1954 was the first to discover the cell-destructive properties of the free radicals, gives us the best general advice on this topic:

(1) Avoid unnecessary calories. In this way one avoids the problem of excess weight and reduces the number of free radicals which arise in connection with the breakdown of foodstuffs.
(2) Avoid those substances which increase the production of free radicals. These include alcohol, tobacco and many of the chemicals which are present in our working environment. It is also dangerous to increase the intake of polyunsaturated vegetable fats, without a corresponding intake of antioxidants.
(3) Increase the intake of antioxidants, which inhibit the destructive effects of the free radicals.

Dr Harman told me recently that he himself takes a daily supplement of 100 μg of selenium, 200 mg of vitamin E and 1.5 g of vitamin C.

5. VITAMINS AND MINERALS PROTECT AGAINST ILLNESS AND PREMATURE AGEING

Our problem is not that we lack sufficient information. The problem is that we have such an excess of the wrong information, that the right information is difficult to find.

Erno Paasilinna

Heart diseases, arthritis, cancer and many other chronic and common diseases derive from the same source: fortuitous mutations, caused largely by free radicals. Under optimum conditions, our cells are protected against the free radicals and lipid peroxidation. There is a complex defence system in the body, in which vitamins, minerals, amino acids and certain enzymes play a central role. This defence is called the antioxidant system. Antioxidants are substances which react chemically with free radicals and render them harmless. At the same time they break the vicious circle which involves the decomposition of fatty acids and proteins, the creation of new free radicals and eventual cell death. Selenium, zinc, manganese, ubiquinone (Q10), vitamins C, E, some of the B vitamins, β-carotene, canthaxanthine, amino acids and certain medicines are effective antioxidants. The principal message in this chapter is that each and every one of us can protect our cells from decomposition by ensuring that we daily receive sufficient antioxidants.

Dietary antioxidants
The food industry adds antioxidants to a wide range of foodstuffs in order to prevent fats from rancidifying. In exactly the same way we can protect our own cells and thus

Examples of diseases which researchers now consider to
be caused or aggravated by free radicals and lipid
peroxidation

Alcoholic liver and heart conditions
Arteriosclerosis
Arthritis
Autoimmune diseases
Cancer
Cataract
Circulation disturbances
Coronary heart disease
Diabetes
Emphysema of the lung
Inflammatory reactions
Liver cirrhosis
Malaria
Multiple sclerosis (MS)
Neuronal lipofuscinosis
Parkinson's disease
Porphyria
Premature ageing
Retinal diseases
Rheumatoid diseases
Senile dementia
Side-effects of medicines

---

prevent cell damage. Our cells also need antioxidants; they are to be found in our
food to a certain extent, but unfortunately none of us get enough of them through our
normal daily food consumption. This is quite clearly borne out by recent nutritional
surveys and blood analyses.

It would appear that the addition of antioxidants to various foodstuffs has been at
least part of the reason for the reduction in the incidence of cancer in the United
States. Antioxidants have been added to most factory-made products there since
1947. In the same period the consumption of fruit and vegetables has increased
significantly — fruit and vegetables are rich in vitamins which function as
antioxidants.

Recently, the incidence of coronary heart disease has fallen in many countries —
in Finland, for example. This is probably due to an increased intake of antioxidants
by the population, combined with a reduced consumption of tobacco. Changes in
eating and living habits have taken place in those countries in which the developments have shown positive results: the use of animal fats has decreased while
consumption of fruit and vegetables has risen. Furthermore, it has become more
popular to take regular exercise, e.g. jogging. High blood pressure is more often
diagnosed and treated at an earlier stage. All these measures contribute to the
prevention of heart and circulatory diseases.

Antioxidants against premature ageing
The most important cells are those of our brains. One of the first signs of ageing is a deterioration in the functioning of the brain, in the memory, for example. Twenty per cent of the weight of the brain is composed of polyunsaturated fatty acids, which easily rancidify; it seems likely that the ageing process in the brain is caused by the rancidification of fatty acids precipitated by the activity of the free radicals. The rest of the nervous system is also vulnerable to rancidification because the nerve ends also contain large quantities of fats. The rancid fat in the nerve cells is known to the researchers as ceroid, melamine or lipofuscin, i.e. age pigment. The oxidation of the nerve ends affects the senses of sight, hearing, smell, taste and touch.

It has been discovered that vitamin C protects the brain and nerve tissues from rancidification. The vitamin C content of spinal fluid is ten times greater than that of other tissue; in brain cells it is one hundred times that of blood. Also the lens of the eye, another target for free radicals, contains much vitamin C. These examples show how nature furnishes the body with the ability to accumulate most vitamin C where its beneficial antioxidant effect is most needed — in the brain and in the eye. We can help nature in its task by ensuring sufficient vitamin C intake. Together with the other antioxidant vitamins and minerals this will contribute to the maintenance of healthy brain function and the avoidance of premature ageing of the brain and to cataract of the eye.

Selenium, together with vitamin E, zinc and other antioxidants, is proving to be a promising weapon in the struggle against premature ageing and the postponement of degenerative diseases: this has been seen from a series of double-blind experiments. These studies, using nursing-home residents in both Finland and Denmark, revealed that the elderly people involved benefitted physically and psychologically from antioxidants. The residents were given an effective dietary supplementation of organic selenium and zinc, plus β-barotene and vitamins A, B6, C and E — i.e. all the natural dietary antioxidants. These studies have received a great deal of attention since they were published and have been referred to in Europe, Asia, Australia and the United States.

Cancer prevention
Antioxidant vitamins and minerals may inhibit the development of cancer. Up until only a few years ago this claim was considered to be ridiculous. However, with the information at our disposal today, it seems not unlikely that this is indeed the case. It is not inconceivable that β-carotene, canthaxanthine, selenium and vitamin C can inhibit the growth of a cancerous tumour.

There is no doubt that selenium protects against cancer. There is a consensus amongst researchers on this point. Experiments with animals have shown that selenium has a successful preventive effect of 50% on average (with variations from 30–85%). In other words, we can prevent half of the incidence of experimentally induced cancer by administering selenium supplementation.

Vitamin A and its precursors, the so-called carotenes, also prevent cancer; this effect is emphasized when it is combined with selenium. An experiment with mice showed that vitamin A and selenium, separately administered, each prevented half of the cases of cancer in those animals which had been exposed to carcinogens. When vitamin A was combined with selenium, however, the preventive effect reached up to 90%; in this latter case, only one mouse in ten contracted cancer. The animals in the control group, which received neither vitamin A nor selenium, all contracted the disease.

In connection with these selenium experiments, it was discovered that vitamin E also has cancer-preventive properties when it, too, is combined with selenium. Several experiments have revealed that 'healthy' people who later contract cancer have a low content of selenium, β-carotene and vitamin E in their serum, compared to those who remain healthy. Large-scale prospective population studies are in progress in the USA and in Finland in an attempt to ascertain the possible protective effects of β-carotene on lung cancer. As yet, not enough is known about the effects of vitamins and minerals where cancer has already developed.

Antidote to heavy metals
Selenium, zinc and vitamin E are antidotes to heavy metals and their cell toxicity. Antioxidants may reduce the harmful effects of these toxic materials to our cells and tissues. Heavy metals are environmental toxins, such as aluminium, cadmium, mercury and lead, which are foreign to our bodies. Vitamin E and selenium can also reduce the harmful and unpleasant side-effects of the various forms of cytostatics which are used in the treatment of cancer.

It has been discovered recently that the amalgam used by dentists to fill the holes in our teeth is the source of minor, and, in some cases major, unwanted side-effects of mercury. Selenium reacts chemically with mercury and can therefore protect us from its toxic properties.

Support to the immune system
Vitamins, minerals, amino acids and essential fatty acids play a vital role in the immune system. The immune defence not only protects the body from viral and bacterial infections, but also from arteriosclerosis, cancer, arthritis, autoimmune disease and many other illnesses. β-carotene vitamins A, B6, C and E, together with the minerals iron, copper, magnesium, selenium and zinc assist the cells in the maintenance of normal immune functions; they also help to re-establish normal immune functions after a period of disturbance or imbalance (illness).

The influence of micronutrients on the immune system can be seen in cases where patients no longer are affected by frequent re-occurrences of or where they are freed from the scourge of the annual attacks of hay fever. The explanation for this is usually that they have received adequate quantities of the dietary factors which support the immune system, either through their diet or from supplementation. In section 6 of this chapter we will look more closely into how this happens.

Improved general health
Doctors generally do not believe that micronutrients can help in the treatment of many different diseases. From their education they have learned that a specific illness should be treated with a specific medicine, while another disease requires another medicine. Doctors do not think there is a panacea, a universal treatment for all problems. Yet, micronutrients may be of great value as adjunct therapy. The explanation is quite simple: in the event of damage, the biochemical functions of the different organs in the body are really rather similar. Vitamins, minerals and

essential amino and fatty acids maintain and repair the normal life-functions of the cells, including their ability to resist disease.

Traditional medical treatment is, in fact, often non-biological. Many drugs block biological receptors in our cells, e.g. beta blockers which are widely used against high blood pressure and ischaemic heart disease or new stomach medicines which block receptors in the gastrointestinal tract. Antibiotics are designed to kill bacteria. However, a considerable proportion of the antibiotics which are prescribed every day are quite superfluous. Young children are often given antibiotics for the common cold, which is caused by a virus, against which antibiotics have no effect. On the contrary, some antibiotics, like tetracycline, can stimulate the formation of dangerous oxygen free radicals in the body. It would be more in tune with nature's intentions if we supported the immune system with micronutrients, which better enable the body itself to fight against infections and inflammations. There are, of course, many patients who already do this — but usually without the doctor's knowledge.

Nutritional therapy usually functions optimally, when one uses substances which support and reinforce each other. Many people — and doctors — have been disappointed when they did not experience any beneficial effects from using one or two substances. Vitamins, minerals, amino acids and essential fatty acids act on each other, but they are also affected by all the other substances which are present in the body — proteins, fats, carbohydrates and hormones, for example. It is of great importance for the defence mechanisms, that there is a balance between all these factors. In Chapter 2, you can read more about nutritional support to medical practice.

6.   THE IMMUNE SYSTEM

We hear and understand only those things that we half know about in advance.

H.H. Thoreau (1817–1862)

Perhaps the term 'immune system' leads us, in the first instance, to think about immunity to, and vaccination against, infectious diseases. Vaccination is, after all, one of the most effective means whereby the body's own defences against communicable diseases can be reinforced.

The human immune defence system gets weaker after the age of 45 years. Deficiencies and disturbances in the immune defence not only increase the risk of contracting allergies and infectious diseases, but also the risk of cardiovascular disorders, rheumatic and other autoimmune diseases, arthritis and cancer. In addition, those changes that follow with age occur earlier if the immune system is already weakened.

In this section, I want to describe the workings of the immune system, and to outline the influence of vitamins, minerals and essential fatty acids on its functioning. A little knowledge of the immune system makes it easier for us to understand the incidence and adjunct nutritional treatment of allergies, infectious diseases, inflam-

mations, cancer, rheumatism, arthritis, heart and circulatory disorders and many other diseases.

Human beings are dependent upon six different types of nutrients in order to be able to grow, develop and maintain good health: proteins, carbohydrates, fats, vitamins, minerals and trace elements — plus water of course. The first three nutrient groups, proteins, carbohydrates and fats, provide energy for growth and the transformation of foodstuffs. Proteins and carbohydrates provide 17 kJ (4 calories) and 38 kJ (9 calories) respectively per gram. About 15% of our daily energy requirement comes from proteins, the remainder is supplied by fats and carbohydrates. The latter three nutrient groups (vitamins, minerals and essential fatty acids) are, however, equally as important in the maintenance of normal bodily functions, although our requirements vary considerably from individual to individual. Our bodily functions are, in part, determined by the balance achieved between the six nutritional factors; if the balance is optimal we remain healthy — if not, then disturbances and imbalances may result in illness.

A well-nourished person has a better resistance to disease than an undernourished person. This is especially noticeable in children and the elderly: an undernourished child or elderly person buckles under much more easily in the event of a simple infection, from which the well-nourished individual rapidly recovers. Malnourishment (deficiency or insufficient intake of one of the six nutrient groups) is the most frequently occurring cause of a failure in the immune system. This problem does not only apply to the famine-hit inhabitants of Third World countries, or to the slum-dwellers in big cities, but to everyone — including both the apparently healthy and those who suffer from chronic illnesses.

A deficiency condition weakens, first and foremost, the cellular immunity (this will be explained below) but it also affects that part of the immune system which produces antibodies. Undernourishment weakens the ability of the white blood corpuscles to destroy viral, bacterial, fungal and cancer cells. Very recently it has been established that essential fatty acids and carbohydrates also play a significant role in the functioning of the immune system.

Vitamins, minerals, essential fatty acids and many amino acids are vital component parts of the immune system. In order to be able to understand what roles they play, it is important to consider the immune system as a whole.

The human immunological defence system is immensely complex. The humoral part of the defence system consists of antibodies, which are manufactured by special white blood corpuscles called B-lymphocytes. The complement system in the serum works together with this defence system. The complement system consists of a number of proteins which circulate with the blood; their job is to transport messages between the antibodies and the antigens (the invading foreign bodies which are to be neutralized). All of the immunologically active cells belong to the other major component of the immune system, the cellular immunity.

We can categorize the immune system by dividing it into four major component parts:

(1) The cellular immunity.
(2) The humoral immune system.
(3) The white blood corpuscles, which attack invading foreign bodies.

(4) The complement system.

These different parts of the system function together interdependently; their effectiveness is dependent upon both genetic characteristics and a range of external environmental factors which we expose ourselves to: environmental chemicals, diet and stimulants.

We often say that a person is immune to a certain disease or allergy. But the immune system has a much more wide-ranging significance for the body. The immune system maintains the metabolic balance, and consequently it protects against disease and premature ageing. In fact, the immune system protects us against all forms of chronic and acute illness.

In the final analysis, almost all diseases and ageing occurs because of a fault or a deficiency in the immune system.

When we die, it is usually the result of a definitive breakdown in the immune system.

Cellular immunity
The most important components of the cellular immunity are macrophages and lymphocytes. Macrophages are a type of white blood corpuscle which circulate in the blood and eat up harmful viral, bacterial and cancer cells. They control the activities of the lymphocytes or white blood corpuscles. They direct the T- and B-cells by transmitting a huge number of different biologically active transmitter substances. It is these signal substances which regulate the functioning of the T- and B-cells, in addition to controlling the process of cell division. The macrophages hereby play an important role in the production of the actual antibodies.

Lymphocytes are the white blood corpuscles, which are formed in the bone marrow and the lymph tissues. There are different types of lymphocytes, each of which have their own function in the immune system. There are two main groups of lymphocytes in the immune system: T-lymphocytes, which are programmed in the thymus for a specific function, and B-lymphocytes, which, on meeting a foreign invader, are transformed into antibody-producing cells which fight the disease provoking invader.

We can divide the lymphocytes into the following groups:

T-lymphocytes
- Regulator T-lymphocytes
  — helper cells
  — suppressor cells
- Effector T-lymphocytes
  — delayed hypersensitivity cells
  — killer cells

B-lymphocytes
— antibody-producing cells
— memory cells

The regulating T-lymphocytes can either strengthen (helper cells) or weaken (suppressor cells) the activity in the other T- and B-lymphocytes. The helper cells assist, amongst other things, in the delayed hypersensitivity response in the skin (eczema, for example), in the rejection of foreign tissue in transplantations, or in the destruction of cells which have been attacked by a virus or which have been transformed into cancer cells.

The B-lymphocytes produce antibodies, also known as immunoglobulins (IgM, IgG, IgA and IgE), on receiving the appropriate instructions from the T-lymphocytes. A failure in this function results in a deficiency of immunoglobulins, which in turn effects a reduced resistance to infectious diseases.

The memory cells remember the immune reactions which have previously taken place in the body, thus facilitating a quicker immune response in the event of a repetition of the same reaction.

For example, if we repeatedly come into contact with allergens in the environment, the production of antibodies against the specific substance occurs more rapidly. It is possible that corresponding regulator B-lymphocytes also exist, but, as with a number of other aspects of the immune system, their existence has not yet been discovered.

Minerals and vitamins also have an influence on the functioning of the so-called neutrophile white blood corpuscles. 60% of all the white blood corpuscles are neutrophiles. These normally live only 4–5 days and this implies that huge quantities of them must constantly be produced. The task of these cells is to locate places in the body which have been damaged or which are under attack by bacteria or fungi. They stick to the invading organisms and consume them. In the course of this process of destruction, lipid peroxides and their toxic end products are given off, which in excessive quantities are harmful to healthy cells (see 1.4). The accompanying table summarizes what doctors are taught in a handbook of immunology about the effects of vitamins and minerals on the cells of the immune system. It is thought-provoking that until very recently nothing was known about the significance of selenium in this context. It is likely that there are other vitamins and minerals, apart from those included in the table, which also have an influence on the immune system. However, we lack sufficient knowledge of them at the present time.

Vitamins

Ever since the earliest discoveries of vitamins, it has been known that they are important for the course of infections and inflammations. Acute inflammation can sometimes be observed in connection with vitamin deficiencies like, for example, vitamin A deficiency, beriberi, pellagra and scurvy.

The quantities of the vitamins A, B6, and C in the body are reduced by viral and bacterial infections like malaria and tuberculosis. The excretion of riboflavin (vitamin B2) through the urine increases in fever conditions. As yet, however, it has not been possible to establish significant alterations in the case of minor infectious diseases. Although we still do not know a great deal about the importance of vitamin intake for infectious diseases, it has been established that vitamins are of great significance in the functioning of the immune system.

The B vitamins, vitamin C and folate are crucial for the normal functioning of the white corpuscles in the event of inflammation. It is also known that the vitamins A, B1, B2, B12, E and folate maintain the defence mechanisms of the macrophages and the T- and B-lymphocytes.

A number of vitamins, A, B6, B12 and folate, for example, participate in the

	T-lymphocytes	B-lymphocytes	Macrophages	Neutrophiles
Vitamins				
Vitamin A	+++	+++		
Thiamin		++		
Riboflavin		++		
Niacin		++		
Pantothenic acid		+++		
Pyridoxine	+++	+++		++
Cyanocobalamin	++	+		
Biotin		+++		
Folic acid	++	+++		
Vitamin C			++	++
Vitamin D		++		++
Vitamin E	++	++	++	
Minerals and trace elements				
Zinc	+++			
Iron	+++	+		+++
Copper			++	++
Magnesium		++		
Selenium	++		++	

+ = minor effect; ++ = medium effect; +++ = strong effect.
Source: Stites, D. P. *et al.*, *Basic and Clinical Immunology*, 1982.

manufacture of proteins and DNA. A deficiency of these vitamins therefore results in reduced resistance to disease. Folate deficiency in pregnant mothers can, for example, cause neurological damage to the foetus.

A reduction in the production of antibodies usually occurs in connection with deficiencies of vitamins A, B6, pantothenate, biotin and folate.

Pyridoxine (vitamin B6) stimulates the production of nucleic acids in the cells. These are essential during growth and in the production of antibodies. A deficiency of vitamin B6 leads to a fall in the number of T- and B-lymphocytes in the blood. Recently, it has been discovered that vitamin B6 deficiency also hinders the production of antibodies. The functioning of the thymus is also dependent upon this vitamin. The thymus programs the T-lymphocytes to be able to recognize and destroy viral, bacterial and cancer cells which it has not previously encountered.

Moreover, it is known that pyridoxine deficiency diminishes the ability of the neutrophile white blood corpuscles to absorb and destroy bacteria. Children, born of mothers with vitamin B6 deficiency, consequently have a smaller thymus and spleen compared with the children of healthy mothers. It is therefore no surprise that newborn children with vitamin deficiencies have a weakened cellular immunity.

Vitamin B6 is also of major importance for the elderly. According to studies by my own and other research teams, there are indications that 25–30% of the elderly are deficient in vitamin B6. This deficiency may actually hamper their mental functions and weaken their resistance to illness and ageing itself.

Vitamin A plays an important role for the T- and B-lymphocytes, and, correspondingly, for the immune defence. Today we know that the vitamins A, C and E protect the skin and the cells of the mucous membranes. A serious deficiency of vitamin A leads to a reduction in the size of the thymus and the spleen, and this is reflected in a heavy reduction in the number of white blood cells, including lymphocytes. Children born with vitamin A deficiency are therefore more receptive to infectious diseases due to the failure of their immune systems. In a UNICEF programme in Indonesia and Africa, child mortality has been significantly reduced by supplementing children and their mothers with vitamin A. It has also proven effective in preventing measles in African children.

But take care: extreme overdosing (more than 10 times the recommended daily allowance) may lead to a weakening of the immune system. Even a lower daily dosage (more than 8000 IU) may be teratogenic and must not be taken during pregnancy.

Together with vitamin A, vitamin C maintains the normal functioning of the neutrophiles and of the skin. The neutrophile white blood corpuscles contain large quantities of vitamin C, which, however, do not work directly on the antibody production or on the activities of the T-lymphocytes. Nevertheless, vitamin C deficiency weakens the delayed hypersensitivity reaction, presumably because the tissues are unable to develop a normal inflammation reaction.

The macrophages also have a high content of vitamin C, especially in the lungs. During a period of infection, the vitamin C content of the white blood corpuscles falls rapidly, reducing their ability to fight viruses and bacteria. When a patient recovers, the vitamin C levels return to normal. It can be worth while to increase one's vitamin C intake at the onset of colds, influenza and other infectious diseases — as some doctors recommended as early as the 1940s.

Vitamin E deficiency causes immunological changes in animals (and presumably also in human beings): the delayed hypersensitivity reaction, lymphocyte activity and antibody production are all weakened. When an animal is given a large dose of vitamin E (5–20 IU per kg per day) the immune system is stimulated, probably because the number of antibody-producing cells increases dramatically. This phenomenon has not been confirmed in human beings, however.

Minerals and trace elements
Zinc deficiency
At the inception of an infection the zinc metabolism undergoes an alteration; zinc is transported to the liver (this is also true of iron, to an even larger extent). The liver begins to produce a protein which contains zinc, and in this way the body is protected. If the infection continues for a longer period, then zinc deficiency can develop, because the elimination of zinc increases through the urine and excreta. At the same time, the intake of zinc often falls because of the patient's lack of appetite.

Protein deficiencies are often associated with iron and zinc deficiencies. Deficiencies of these very minerals provide an explanation for many disturbances in the immune system. Many of the functions of the cells are dependent upon zinc: Vitamin A production, for example, requires zinc in order to be able to transform retinol to retinal. Energy production in the cells also requires zinc; this vital mineral is also essential in the formation of nucleic acids (RNA and DNA), while hundreds of different enzymes are also dependent on zinc, for example copper–zinc superoxide dismutase (CuZn SOD) (see pp. 179, 182–3).

Zinc improves the functioning of all cells in the cell division phase, including the cells in the immune system. Zinc is therefore vital to the maturation of the lymphocytes. Zinc deficiency weakens the immune functions of the cells, especially those of the helper-T-lymphocytes and the so-called killer cells.

As is the case with vitamin A deficiency, zinc deficiency leads to an atrophying of the thymus; this in turn results in a fall in the production of those hormones which the thymus supplies to the body, while antibody production, which is dependent on the thymus, increases. In addition, the number of so-called nul-cells in the blood increases. Nul-cells are white blood corpuscles which are neither T- nor B-lymphocytes, and therefore have no specific immunological function.

While nutritional zinc intake (15–30 mg/day) usually corrects zinc deficiencies, massive overdosing with zinc, i.e. more than 200 mg/day, can lead to a weakening of the immune system.

Iron Deficiency

Iron deficiency is known to be associated with infectious diseases, but it is difficult to ascertain what is cause and what is effect. The optimal functioning of the lymph tissues demands iron; iron is also necessary for the formation of certain enzymes and for the ability of the neutrophiles to digest and neutralize foreign substances in the blood.

Transferrin and lactoferrin are enzyme proteins which form compounds with iron and thereby inhibit the growth of bacteria. These enzymes absorb all the available iron, making it impossible for the bacteria to reproduce (owing to the lack of iron).

However, not all forms of iron deficiency can be cured with the aid of iron supplementation. When iron is administered to a person with a long-term infection or inflammation, e.g. rheumatoid arthritis, the anaemia is not cured. In fact, it has been shown that when a patient is given an injection of iron, the iron is stored in the serum and in the tissues without having any effect on the anaemia. In such a situation the patient can actually become even more receptive to new infections, because the transferrin in the serum is totally blocked by iron. Very recently, it has been shown that individual cells of patients suffering from rheumatoid arthritis may contain an overload of iron, although their haemoglobin values are low. In this instance there is no iron deficiency in the cells, rather the number of red cells (haematocrit) is pathologically low. Here iron supplementation may not help the patient; on the contrary, it can eventually be directly harmful.

Iron deficiency damages the cellular immunity, while the delayed hypersensitivity reaction and the activity of the T-lymphocytes are partially obstructed. Similarly, the ability of the neutrophiles to eliminate bacteria deteriorates, because important enzymes, which are dependent on iron, are no longer able to function. In some cases the production of antibodies is also seen to drop.

These disruptions to the immune system already occur when the iron intake is reduced by only 10%. It can be, that a deficiency of folate is an additional explanatory factor in this phenomenon.

Copper, magnesium and selenium deficiencies

An acute or more long-term infection increases the *copper* content of serum, because the production of ceruloplasmin (a copper-rich protein) increases in the liver. This would appear to be one of the body's defence mechanisms against infections, but it has not been fully explained as yet. After the infection has passed, the copper content of the blood normalizes in the space of a few weeks. It has been established, however, that copper deficiency has a harmful effect on a component of the activity of the immune system, the so-called reticulo-endothelial system.

Magnesium deficiency weakens the production of antibodies and leads to atrophying of the thymus. Magnesium and calcium are necessary to the normal functioning of the cell membrane metabolism. During an infection, the magnesium content of the blood falls; this could be due to a dilution of the serum, because the illness releases tissue fluids into the serum.

It has been discovered, in experiments with volunteers, that the body loses large quantities of calcium and magnesium during an infection, mainly from the cells. Magnesium has a beneficial effect on the formation of complement in the serum, a part of the human immune defence system. Magnesium supplementation is vital when large quantities of the mineral are lost, in the event, for example, of serious burns or post-operational complications. Magnesium would also appear to be beneficial for the prevention of allergies — and possibly also for common infections.

Selenium deficiency weakens the cellular immunity, especially when this occurs simultaneously with a low level of vitamin E. The ability of the neutrophile white blood corpuscles to destroy antigens ceases and the activity of the T-lymphocytes is weakened. These changes can be connected with the antioxidant effect of selenium, because the rancidification of the fatty acids in the cell membranes also increases. Simultaneous mild deficiency of selenium and vitamin E has been associated with an excess risk of human cancer, as recently illustrated in prospective epidemiological studies in Finland (see p. 94).

The humoral immune system

Cortisone, thymulin and interleukin-2 (IL-2) belong to what is called the hormonal immune system. IL-2 production is highly dependent on the individual's zinc status. A marginal zinc deficiency, a common complaint amongst the elderly, causes many forms of disturbances in the immune system, i.e. weakened resistance to diseases and to the ageing process.

The activity of the thymulin is likewise dependent on zinc. It is when faced with such situations, that one perhaps should consider zinc supplementation: that is, when zinc deficiency is established in cases of allergies, rheumatic and arthritic complaints, inflammations, and ageing.

7.   RECOMMENDED DIETARY ALLOWANCES

> *Actually, the individual has some knowledge, even if it is only very little; the*
> *more knowledge he possesses, the more he begins to doubt*
>   J. W. von Goethe (1749–1832)

The central question in the discussion about micronutrients is, of course, whether or

not we get enough of them from our daily diet. Dietary experts and many doctors tend to be of the opinion that a varied diet is sufficient, and that food supplements are therefore unnecessary. The guidelines for Recommended Dietary Allowances (RDA) are usually based on American norms. It is on the basis of this that the intake of most vitamins and minerals is considered to be sufficient for the population at large.

On what then, are these 'recommended dietary allowances' founded? Are they aimed at the sick, or at healthy people? Do they refer to average population groups or to individuals? Are the recommended figures reliable and why are the experts so divergent in their opinions? I will try to answer some of these questions below.

The US recommendations which are in use today stem from 1943, although they have been updated at five-year intervals since then (most recently in 1989). *The RDAs lay down the daily intakes of proteins, vitamins and minerals which are required for the maintenance of good health.* The recommendations have an inbuilt safety margin. In this connection it is important that we point out two things: firstly, the recommendations apply to the *population* as a whole, not to the individual, and, secondly, they apply to an *average healthy population*. In reality, not a great deal is known about the needs of *the individual*, neither of the healthy nor the sick.

The individual RDAs make clearly expressed exceptions of premature babies, pregnant and nursing mothers, all those over 50 years of age, people with metabolic disorders, those who suffer from infectious or chronic diseases and those who suffer from some illness which requires that they receive permanent medication. There are special RDAs for pregnant and nursing mothers, for men, women and children, but none for the over-fifties, smokers, alcohol users or people who are ill.

Very few people, and this includes most doctors, really know what the recommendations are based on, or precisely how these figures are arrived at. When the original recommendations were made, only those quantities which were required to prevent actual deficiency symptoms were taken into consideration. No clinical investigations were undertaken to ascertain any further influence of vitamins and minerals on health. When the directives were first issued in 1943, it was quite simply believed that the amounts which were contained in the average American daily meals must be sufficient. No consideration was given, for example, to those amounts of left-overs which ended up in the rubbish bin. In these recommendations, only those quantities were considered which were needed in order to prevent well-known deficiency symptoms; *no consideration was taken in 1943, or later, of the prevention of cancer, cardiovascular diseases, rheumatic or arthritic complaints or immunological effects.*

Simple measurements of the vitamin and mineral content of various foodstuffs tell us little about their absorption and effects in the organism, and nothing whatsoever about the mineral balance in the individual. Actually, there are considerable variations in the bio-availability of, say, selenium from various sources, e.g. wheat, fish, meat and mushrooms. *Bio-availability* refers to the degree to which the individual organism is able to absorb and utilize vitamins and minerals. Moreover, there are individual differences which result from varying genetic characteristics, cigarette smoking, alcohol consumption, and use of medicines. In the RDAs, no consideration is given to the fact that in some cases substances work against each other, or are affected by other foodstuffs. When one thinks of how differently healthy people view the question of their daily requirements, it is no surprise that the confusion is even greater when it comes to the evaluation of the needs of the sick or of special high risk groups in the population. The confusion is first and foremost due to the unreliability of our basic knowledge.

Let's take an example, vitamin E. The RDA for vitamin E is 10 IU or milligrams per day. However, a daily dose of 100 mg (IU) was recently suggested, in an article published in the *American Journal of Clinical Nutrition*, in order to ensure a minimum vitamin E concentration of 30 micromoles per litre human blood serum, which was considered to be a safe level for protection from ischaemic heart disease and cancer. Several renowned scientists, who have been investigating the biochemical and medical effects of vitamin E against free radical induced cell damage, themselves take daily as much as 200–400 mg (IU). A Japanese professor of nutrition, who takes 400 mg per day himself, stated that the heavy environmental pollution has quite simply made the traditional RDAs obsolete. There is good reason to believe that vitamin and mineral requirements are greater for the sick than for the healthy. Apart from this, everybody who lives from an unvaried diet may need food supplements. The lack of a varied diet can be due to, for example, food allergies like gluten intolerance or lactose intolerance, gastrointestinal illnesses, general poor health or loss of appetite. This is often the case with the elderly, or others who can no longer be bothered to cook for themselves.

Vitamin and mineral requirements are evidently greater than RDAs for those who have a high alcohol or coffee consumption, those who use medicines regularly (diuretics, contraceptive pills, hormones, antibiotics or cytostatics), or those who are undergoing radiation treatment. Recent studies have revealed that smoking also affects the absorption capacity and blood levels of at least β-carotene, vitamin C, selenium and zinc, of which smokers ought to have a higher intake than the RDAs. It is worth while noting that the US National Cancer Institute currently sponsors several β-carotene supplementation studies where the daily supplement is at least 15 mg. This is about seven times the actual daily intake in average meals and twice as much as the RDA for vitamin A (in retinol equivalents).

Most chronic and long-term illnesses increase the excretion of minerals; in fact, one can talk in terms of disease as something which 'consumes' vitamins and minerals. Diabetes, kidney diseases, long-term inflammations and infections, rheumatic and arthritic complaints and cancer are all examples of such consuming diseases. Most chronic diseases and most medicaments stimulate an increase in the production of lipid peroxides, so for the sick person, it is even more essential to increase the intake of antioxidants than for the average healthy individual. In fact, one of the crucial questions in the treatment of cancer patients is the treatment and prevention of undernourishment. In this context vitamins, minerals, and amino and fatty acids are of immense importance.

In my opinion, the RDAs cannot be regarded as the optimal intake of many vitamins and minerals.

8. QUESTIONS OF EXPERIMENTAL METHODOLOGY

Science has the first word on everything and the last word on nothing.
                                                                Victor Hugo (1802–1885)

We do not know what the exact individual requirements for vitamins and minerals

are, and it is even more difficult to define the needs of a person who is chronically ill. Our knowledge about vitamin and mineral requirements is so lacking, that it is difficult to investigate into what effects they have in the prevention and treatment of illness. To date many of the investigations have been so-called 'nul-experiments', i.e. they have failed to provide us with any new knowledge. In this section I want to outline the problems of research into this field.

Vitamin and mineral research can be traced far back into the past. Until the 18th century, one of the worst diseases confronting long-distance sailors was scurvy; at this time it was discovered in the British Navy that this could easily be cured by consuming lemon juice (hence the American slang for the British, 'Limey'). Henceforth the consumption of lemon juice became obligatory. Today we know that scurvy is caused by vitamin C deficiency, and that lemons are rich in vitamin C. This is an example of how we may find effective means for the prevention of an illness (scurvy), even if we do not know the direct cause (vitamin C deficiency).

A more recent example relates to pernicious anaemia: up until the middle of this century, everyone who contracted the disease died. It was then discovered that fresh, raw liver could keep sufferers of pernicious anaemia alive. Today we know that it is the vitamin B12 in liver which is responsible for this 'miracle'. In these cases the causal context has been established, but it is often difficult to arrive at a clear conclusion about which specific substance or factor is responsible for a specific effect.

Animal experiments

Vitamin research started in earnest early in our own century with animal experiments on special diets. The experimenters excluded first one, then later another, foodstuff from the diets of the animals. Through this method it was possible to detect deficiency symptoms. It was later confirmed that the deficiency symptoms were associated with certain biochemical compounds, and, when it became possible to isolate these compounds, they were given the name vitamins (the 'amines of life').

It was in the course of this kind of vitamin research that Dr **Klaus Schwartz** discovered, in the USA in 1957, that selenium deficiency could lead to liver necrosis in rats. This was the starting point for the intensive research on selenium which is now proceeding all over the world.

Most of the mineral research today must necessarily be conducted with the aid of experimental animals, for example in the case of cancer prevention studies. The results are then used to form conclusions, insofar as these relate to human beings. It can, of course, be difficult to adapt the findings of animal experiments as such to human beings, because the physical structure and the metabolic processes of animals are often very different from those of human beings.

On the other hand, it is of course quite out of the question, in the long term, to administer special vitamin- and mineral-deficient diets to humans. Neither is it possible to administer toxic vitamin and mineral dosages to humans just in order to see what will happen. Similarly, we cannot possibly transfer cancerous tumours to humans and then attempt to treat and cure them with vitamins and minerals. Consequently, the lion's share of our knowledge about vitamins and minerals stems from animal experiments.

Experiments involving human beings

The most old-fashioned method of publicizing new discoveries is to describe one or more cases of the observation of a new phenomenon. This type of *case study*, once the dominant one in medicine, is no longer valued very highly. There are, of course, exceptions, in the event of revolutionary or incontrovertible finds, or cases of specific serious side-effects.

Another mode of presenting new discoveries is to collect a whole group of patients who share the same symptoms and the same effective treatment. In the philosophy of science this method is called *induction* (inductive type of conclusion). It should be borne in mind that *a greater part of current medicinal therapy is still based on the inductive accumulation of evidence*. These types of investigation, however, are usually rejected today by medical journals because of the lack of a corresponding group of controls. The purpose of a *control group* is to show what would have happened if the original group had not been treated as they were.

It is even more important to have a control group when the effects of treatment with a specific new medicine are being investigated; it is then possible to compare the results with those of a group who did not receive the medicine or a who were given another kind of medicine, or placebo (a pill with no pharmacological effect).

There are two types of comparative study:

(1) Case-control (patients/healthy controls).
(2) Cohort studies and follow-up studies

In a *case-control study*, the patients (cases) are chosen from those who are suffering from the particular disease which one wishes to investigate, while the control group consists of healthy people, or those who are suffering from a disease other than the one which is being investigated. After the selection of subjects, one can investigate the desirable or undesirable effects of, for example, β-carotene intake, smoking habits, coffee consumption, dietary fibres, etc., on the outcome, i.e. the disease.

An example: We collect a group of lung cancer patients and establish that they have had a lower intake of β-carotene (a precursor to vitamin A), or that they have lower serum levels of β-carotene, than a group of healthy controls who have consumed more β-carotene. We can therefore conclude that low β-carotene intake is a risk factor for cancer.

This form of investigation (case-control) is advantageous from the point of view of time. The results can be achieved relatively quickly after the commencement of the investigation. This kind of study is called *retrospective* because it deals with what has happened in the past. It can, however, be a serious problem to acquire reliable information about the object of the study (consumption of foodstuffs, for example), while it is also difficult to define precisely something like carotene intake. Because of the several confounding aspects, *the case-control design can not prove cause and effect*.

A *follow-up study* is *prospective*, i.e. it is aimed at observing future developments. At the beginning of the investigation there are two groups, both of which are free of the disease which is to be investigated. Thereafter, one of the groups is, for example, given a specific medicine, while the control group receives no medicine. During the course of the study both groups are observed closely in order to ascertain whether or not the group which has been under treatment has received any beneficial effects, compared to the group which has not received treatment. This type of study design is generally preferred because it yields evidence for or against a hypothesis. In the philosophy of science this is called the *hypothetic-deductive method*.

To date, more than 30 major prospective vitamin and mineral studies have been carried out in which blood tests were taken from healthy individuals and the condition of their health was observed over a period of time, with reference, for example, to the development of cancer. Such studies have revealed that many of the later cancer patients already had (as 'healthy' individuals) low blood levels of, amongst other things, vitamin E and selenium from the inception of the investigation. The control group is necessary in order to ensure that all the other factors in the environment and in living habits, which can exercise an influence on the disease during the course of the investigation, are controlled or excluded as far as is reasonable.

A major problem with this type of investigation is that it must necessarily be *long-term*, sometimes lasting several years or even decades. Such a study inevitably becomes very expensive. A shorter follow-up period for a study on a specific preventive treatment for cancer for example, would be worthless, because cancer takes so many years to develop. For example, in a Finnish cohort comprising 12 000 persons, a total of 56 cases of cancer developed during four years of follow-up. In another Finnish survey, 343 cases of cancer were diagnosed over a period of ten years from a group consisting of 15 000 women.

Another problem is *the size* of the two groups which are to be compared. The rarer the disease is, the larger the groups have to be. Let us imagine, for example, that we want to investigate the influence of dietary fibre on cancer of the large intestine, something which perhaps occurs in one out of one hundred individuals in the course of ten years. In order to prove that a case of cancer of the large intestine has been prevented by an increased intake of dietary fibre, there would need to be at least 1000 people in each of the two groups. Nine cases of the cancer should therefore occur in the treatment group, as opposed to ten cases in the control group. However, coincidences can produce a result where 9, 10 or 11 cases arise in each of the groups; consequently dietary fibres have no apparent effect on intestinal cancer. In medical language one would say that there is no visible *statistical significance* for the beneficial influence of dietary fibres on intestinal cancer. But this interpretation is incorrect. In order really to be able to prove an effect in this sort of study, there should perhaps be 100 000 participants, and a follow-up period of 20, instead of only 10, years.

It is therefore hardly surprising that there are so many *negative experimental results* in the field of dietary research. Effects are not evident on the basis of such small experimental groups and yet much too wide-ranging negative conclusions are drawn from the same research. In such cases, it would be more in accordance with the truth to admit that the results were '*non-positive*', i.e. that it is a question of a *nul-*

study which neither confirms nor refutes the hypothesis. Moreover, if the disease in question follows a more unpredictable course, as is the case with, for example, disseminated multiple sclerosis, then the results will be unreliable.

The question of *expense*, as we have mentioned, is also a problem. If, for example, we wanted to conduct an investigation into the potential benefits of treating disseminated sclerosis with γ-linolenic acid (GLA), it would require several hundred patients and a follow-up period of at least five years. There would need to be a control group, and placebo capsules resembling the yellow GLA capsules would need to be manufactured. The cost of such a study would run into several tens of thousands of pounds. On the grounds of cost alone, it would be difficult to carry out such an investigation without the financial and personnel facilities of the larger multinational pharmaceutical firms or research institutions.

The biggest problem associated with all population studies, however, is to find *an appropriate control group*. It is of the utmost importance that the treatment group and the control group are identical as regards to age, sex, coffee, tobacco and alcohol consumption, work, social status, housing conditions and other factors which might influence the outcome under investigation, e.g. the incidence of cancer.

Still another problem is *compliance*. In research experiments it can be difficult to motivate the participants to take pills over months or even years, especially when the patient is aware that there is a 50% chance that he has landed in the placebo group. (Read more about placebo in the next section.)

Vitamin and mineral studies can be divided into two main groups. In the first group one investigates the connection between a specific illness and a specific deficiency symptom; in the other group one studies the effects of vitamin and mineral treatment on a disease. Generally speaking, it is easier to investigate deficiency conditions. For example, we can measure the vitamin and selenium content in the blood of heart attack patients and compare them with those of healthy controls. If we detect lower blood values for these micronutrients in the patients than in the control group, we can establish a statistical correlation between selenium deficiencies and ischaemic heart disease. Whether or not this correlation really shows *cause and effect* is quite another matter. Is it just a matter of coincidence, or is there really a causal connection? Interpretation of such a statistically significant result is difficult — it could be that numerous other confounding factors are of more importance.

Most laymen probably know the old joke about the statistical link between the rate of ice cream sales and drowning accidents on a hot summer's day. I should perhaps point out, in this context, that this is just an example, and that the statistical evidence for the relationship between low selenium status and ischaemic heart disease is, in fact, well documented, as is the evidence for selenium deficiency's correlation with cancer. These epidemiological discoveries have given researchers a lot of headaches, and their statements to the public have been on the over-cautious side because epidemiology does not yield 100% certainty.

In an attempt to clarify the relationship between β-carotene and cancer, two major studies are under way at the present time — one in the USA and one in Finland. In the USA 22 000 doctors are taking 50 mg of β-carotene, or a placebo (pills with no pharmacologic effect) every other day. In the Finnish study 30 000 smokers take 20 mg of β-carotene or a placebo daily. In due time we will know more about the effects of β-carotene on lung cancer in smokers.

9.   PLACEBO

The doctor has only one task: to heal the sick, and if fortune is with him, it is of no consequence which means he uses.

Hippocrates (460–370 BC)

Placebo **is a Latin expression and means 'I seek comfort'. The object of placebo treatment is quite simply to bring comfort to the patient. The word placebo is, however, used in two different contexts: placebo effect and placebo medicine. The placebo effect is seen in both placebo medicine (pills with no active ingredients) and in 'real' medicine.**

By placebo medicine, we mean a kind of ineffective 'cheat' medicine which lacks any form of pharmacologic effect. Nevertheless, it is used both in experimental and treatment situations. If one administers placebo medicine, and if the patient believes that it is real medicine, in many cases this belief is sufficient to precipitate a cure.

The expression 'placebo effect' is often used as a pejorative to describe pharmacologically ineffective treatment. Many studies have proved beyond all doubt, however, that placebo does in fact have an effect. If one gives an injection, and maintains that is morphine, a substantial proportion of patients will experience an easing of the pain, despite the fact that one has only administered a salt solution.

Studies of stomach ulcer medicines have revealed that stomach ulcers were healed in 20–40% of patients who received placebo medicine. A study has been published recently, in which cimetidine (e.g. Tagamet) was compared with acid-neutralizing medicine and placebo. The investigation was carried out over a period of 12 weeks. Stomach ulcers had been diagnosed for all the patients in advance, using gastroscopy (an investigation where the doctor, by means of a flexible tube, can observe the mucous membranes of the stomach and the duodenum). The effects of the treatment were evaluated after a further gastroscopy.

Below, the results of the treatment for the different groups are listed as percentages of success.

Period of investigation (weeks)	Percentage success		
	Cimetidine	Acid-neutralizing	Placebo
4	53	38	36
8	86	70	58
12	89	84	70

In this study all the patients had ulcers of approximately the same degree of severity. The results show quite clearly that cimetidine is more effective than either acid-neutralizing or placebo medicine — the results are statistically significant. But we can read the results in another way: placebo would appear to be a perfectly good medicine for most stomach ulcers. In the course of three months, 70% of the placebo patients were cured. We may conclude that the pharmacological effect of cimetidine accounts for 19% of the patients; the rest would probably have been cured by placebo as well. This is most certainly not the way in which the new medicine was introduced to doctors.

It is estimated that about a third of all patients can be successfully treated with placebo, covering a wide range of complaints. And the effect is usually greater when the medicine is more expensive (i.e. the placebo). The doctor's attitude to the patient and the treatment also plays a major role: if the doctor is in doubt about the effectiveness of the treatment, the patient often picks up on this, and this probably reduces the effects.

Under no circumstances should we consider the placebo effect as something negative: the patient with psychosomatic (stress) symptoms seeks the alleviation of his ailment, and if this is achieved with the aid of safe placebo medicine, then all is well. If, on the other hand, we are confronted with a case of cancer, or some other correspondingly serious illness, then it is clear that such symptomatic treatment does not suffice.

Since *all forms of medicine have a placebo effect*, all new medicines are tested with the aid of a control group who are given placebo. This kind of test is usually carried out double-blind. If the two groups are furthermore exchanged midway through the experiment then it is called a cross-over. The cross-over-double-blind study is regarded as the most reliable, and respected medical journals publish more or less only this kind of study. A reliable, clinical investigation demands, in addition, that there are sizeable patient and control groups before the results can be statistically significant.

The placebo effect varies from illness to illness. Serious, advanced diseases like cancer, severe asthma, painful arthritis, psoriasis and severe angina pectoris do not react to placebo treatment to the same degree as the psychosomatic diseases.

Although placebo is medicine without a pharmacologic effect, it still gives rise to side-effects in quite a few experimental participants. The most common side-effects are fatigue, drowsiness, insomnia, headaches, general discomfort and dryness of the mouth. This is known as the *nocebo* effect (i.e. contrast to placebo effect).

Of course the placebo effect plays a role in nutritional medicine and applies to nutritional treatment just as it does to all other medical treatment. If, after a course of treatment (or of any other form of treatment), the patient feels better, it still does not prove that we witness a 'pharmacologic' treatment effect. The illness might have been cured without treatment, or another course of treatment might have been administered concurrently, while the placebo (or nocebo) effect can never be totally dismissed.

Those who have confidence in natural medicine often choose, consciously or unconsciously, to forget all about the placebo effect. Their opponents, on the other hand, vociferously demand pharmacological proof for the effects of natural medicines. Reliable research results would, however, demand that long-term double-blind studies were carried out with the participation of very large patient and control groups. In the great majority of cases it would be impossible to undertake such studies, mostly on the grounds of immense expense.

There are many other problems associated with medical research. The main purpose of this chapter, however, has been to give the reader an impression of the difficulties connected with nutritional medicine, including vitamins and minerals. This provides a better background for understanding the contemporary debate about these treatments; a debate which from time to time has been extremely heated.

LITERATURE

Blake, D. R., Allen, R. E., and Lunec, J. Free radicals in biological systems — a review orientated to inflammatory processes. *Br. Med. Bull.* **42** No. 2, (1987) 371–385.

Davies, S. and Stewart, A. *Nutritional Medicine — Drug Free Guide of Better Family Health.* Pan Books Ltd, London, 1987. p. 543.

Halliwell, B. and Gutteridge, J. M. *Free radicals in biology and medicine.* Clarendon Press, Oxford, 1987.

Hurley, L. S., Keen, C. L., Lönnerdahl, B. and Rucker, R. B. (eds). *Trace elements in man and animals* **6.** Plenum Press, New York — London, 1988, p. 724.

Hayaishi, O., Niki, E., Kondo, M. and Yoshiwaka, T. (eds). *Medical, biochemical and chemical aspects of free radicals.* Vols 1 & 2. Elsevier, Amsterdam — New York — Oxford — Tokyo, 1989, p. 1559.

Pryor, W. A. The Free Radical theory of ageing revised: A critique and a suggested disease-specific theory. In: Warner, H. R. *et al.* (eds) *Modern Theories of Ageing.* Raven Press, New York, 1987, pp. 89–112.

Slater, T., Cheeseman, K., and Davies, M. *et al.* Free radical mechanisms in relation to tissue injury. *Proc. Nutr. Soc.* **46** (1987) 1–12.

Southern, P. A. and Powis, G. Free radicals in medicine I. Chemical nature and biological reactions. *Mayo Clin. Proc.* **63** (1988) 381–389.

Southern, P. A. and Powis, G. Free radicals in medicine II. Involvement in human disease. *Mayo Clin. Proc.* **63** (1988) 390–408.

Stites, D. P., Stobo, J. D., Fuderberg, H. H., and Wells, J. W. *Basic and clinical immunology.* 4th edition Lange Medical Publications, Los Altos, California, 1982, p. 775.

Takeda, K., Shimada, Y., Amano, M., Sakai, T., Okada, T. and Yoshiya, I. Plasma lipid peroxides and α-tocopherol in critically ill patients. *Critical Care Magazine* No. 12, 1984, pp. 957–959.

Tappel, A. L. Free radical lipid peroxidation damage and its inhibition by vitamin E and selenium. *Fed. Proc.* **24** (1965) 73–78.

Tolonen, M. Finnish studies on antioxidants with special reference to cancer, cardiovascular diseases and ageing. *Int. Clin. Nutr. Rev.* **9** 2 (1989) 68–75.

Recommended Dietary Allowances (RDA). National Academy of Sciences. 10th edition Washington, 1989.

2

Micronutrient therapy

1. WHAT IS MICRONUTRIENT THERAPY?

Natura curat, medicus adjuvat. (Nature cures, the doctor helps.)
Latin proverb

Nature is not a temple, it is a workshop in which man works.
Ivan Turgenev (1818–1883)

Many vitamins and minerals are antioxidants which inhibit or repair cell damage. In addition, in concert with essential amino and fatty acids, they regulate the prostaglandin metabolism and support the immune system. This explains why it is possible to improve human resistance to, amongst other things, allergies, asthma, skin diseases, infections, rheumatic and arthritic complaints, cardiovascular diseases and cancer.

Nutritional therapy is distinguished from more conventional medicine in that it is not only directed at the disease itself and the accompanying symptoms; it is also aimed at improving the patient's natural defence mechanisms, including the immune system. The target is the person rather than the disease, in contrast to pharmacological-specific medication. The nutritional approach can be regarded as less specific and more 'holistic', in that the aim is to affect every one of the patient's trillions of cells.

Moreover, we are dealing with a form of treatment which is 'closer to nature' because it does not involve the use of any substances which are foreign to the body. It is safe to employ vitamins, minerals and essential amino and fatty acids under professional supervision. Why then has the debate about these substances become so heated?

Vitamins, minerals and essential amino and fatty acids work closely together in the body like an orchestra to reinforce each other's physiological effects. The intention of the treatment is to resuscitate the patient's own resistance to disease, and compared with most traditional medicinal treatment, this has to be extended over a much longer period of time. *Micronutrient treatment is not an alternative therapy*; it is based on scientific biochemistry and it is often used as a supplement to support the effect of another medicinal treatment. Micronutrient therapy combines well with more traditional treatments in cases of, for example, allergies, eczema, intestinal diseases, cancer, rheumatism and arthritis and heart diseases.

Vitamin and mineral treatments, together with essential fatty acids and amino acids, have increased in popularity in step with the growing awareness of selenium and other 'new' trace elements. Vitamin supplements are widely prescribed by doctors in Germany, Austria, Switzerland and France, while in countries with a strong Anglo-Saxon influence the medical establishment remains more reticent. However, it has become more and more of a rarity to hear of doctors who condemn such treatment out of hand. This is first and foremost due to the constant expansion of our knowledge in this field and to the frequent discoveries of 'new' effects of vitamins, minerals and essential amino and fatty acids. Their popularity is also connected with the remarkable results which have been achieved and to the increasing desire on the part of patient to be treated more 'in accordance with nature'. This desire is fulfilled in that the patient is treated solely with substances with which the body is already familiar instead of with synthetic foreign chemicals.

It is best for the patient if the treatment is based on an assessment of that patient's vitamin, mineral and fatty acid blood levels by a doctor who has experience in the field. The amino acids are usually determined in the urine. The doctor can then prescribe the appropriate micronutrients in accordance with his own examination and assessment of the patient's symptoms and the results of laboratory tests. It is vital that the treatment is organized so as to ensure that the patient receives all of the nutrients in which he or she is deficient. Treatment with only one or two substances often leads to disappointing results, as was the case in a recent study in Denmark, where an attempt was made to treat chronic rheumatoid arthritis with selenium alone. Supplementation treatment of chronic diseases, as we have mentioned, often extends over many months or years, and sometimes for the rest of the patient's lifetime.

In the next section I want to tell more about mineral experiments and give some examples of diseases which I have treated with vitamins, minerals and essential fatty acids myself, often as a supplement to another course of treatment. This largely involves descriptions of individual cases and the reactions of specific patients to the treatment. These cases are not meant as scientific evidence but there are a number of interesting cases in which both doctor and patient have been able to share the satisfaction of positive results.

It is important to establish that micronutrient treatment is completely safe for the patient. In my opinion, there is no reason why a patient who is suffering from a serious or long-term illness should not try a treatment which is perfectly safe — even if its effects have not yet been proved fully.

In early 1988 the Finnish health authorities and the Finnish Medical Association accepted that vitamin and mineral treatment should no longer be classified as 'alternative medicine'; it is now classified as officially accepted medicine. There would also appear to be great interest in the field amongst doctors in Denmark, Sweden, Norway and the UK: the halls were full to bursting point at my recent lectures on the subject in Copenhagen, Stockholm, Oslo, Vienna and Newcastle upon Tyne.

2. MICRONUTRIENT ANALYSES

Everybody is ignorant, but about different things.

Will Rogers (1879–1935)

Ignorance solves no problems.

Benjamin Disraeli (1804–1881)

It is possible to measure the intake of vital micronutrients from diets, but the actual content in the body is best measured from an analysis of the blood or other tissue. This kind of test not only provides information about possible deficiency conditions, it also enables doctors to monitor the course of the treatment. Blood tests ensure, moreover, that the patient does not receive too much of any particular vitamin or mineral. Professor Fred Gey of WHO has recently published risk thresholds for antioxidant blood levels in relationship to cancer and heart disease prevention (see page 98).

Vitamin, mineral, amino and fatty acid intake is usually measured in two different ways. The first method is most often employed by nutritionists and dietitians. A *typical daily diet* is put together and the content of, for example, selenium and chromium is measured. Another, better method is the so-called *double-portion* technique: a perfect copy is made of the actual diet under study, and the content is then analysed in the laboratory. The drawback with these methods is that it only takes into account the quantities of the various micronutrients which are present on the plate, leaving us unaware of what the individual actually consumes. No consideration can be given, for example, to individual differences in absorption of the nutrients.

Neither of these two methods are able to shed much light on the actual micronutrient content in the patient's blood or tissues. It is, however, possible to measure the blood and urine content and to carry out other accurate forms of analysis. A more realistic assessment of the *bio-availability* of the substance and of the patient's mineral balance is then possible. The capacity to absorb, for example, fat-soluble vitamins from diet (or from vitamin pills) varies considerably from one individual to another.

Only very tiny quantities of minerals and trace elements are present in the blood. In the case of selenium, chromium and manganese the figures are measured in terms of micrograms, while the largest quantities, in the case of zinc, for example, are measured in terms of milligrams. A microgram is a one thousandth of a milligram and a one millionth of a gram. It can of course be difficult to measure such minuscule quantities. In fact, it is only in recent years that equipment has been developed which is capable of producing reliable measurements to this degree of precision. As far as selenium and chromium are concerned, this has only become possible in the 1980s. Even today, it can still be difficult for many laboratories to produce accurate measurements of these minerals. It is also possible to determine the content of all vitamins in the serum, but here again, results may vary from one laboratory to another.

Serum tests have, however, been criticized because it has been discovered that there can be serious mineral deficiencies in the *cells* even though mineral content in the serum is perfectly normal. A serum test is therefore not always an accurate pointer to a genuine deficiency condition. It is common knowledge amongst doctors that measurements of potassium and magnesium in serum are far from optimal but until now there has been no better method to estimate the potassium and magnesium balance.

New developments in laboratory equipment are, however, now making it possible to undertake mineral analyses of blood cells, which provide far more accurate information about the patient's mineral balance. Two Swedish researchers, Dr **Ulf Lindh** of Uppsala University and Dr **Erland Johansson** from the Gustav Werner Institute in Uppsala, have recently developed a microanalysis, which can determine the mineral content of the *individual cell*. The method, called microPIXE, is extremely precise and can establish simultaneously the content of several different minerals within the cell. They have so far been able to detect characteristic alterations in the cell micronutrient levels in cases of rheumatoid arthritis, breast cancer and juvenile ceroid lipofuscinosis. There is no doubt that we will hear a great deal more about these Swedish experiments in the future.

Vitamins and minerals operate in the body through a number of enzymes and it is sometimes possible to determine the nutritional status, through *functional tests*, e.g. by measuring the enzyme activity. The bio-availability of selenium, for example, is reflected to a certain extent in the activity of the enzyme glutathione peroxidase. Other functional tests, relevant to lipid peroxidation, are analysis of exhaled pentane and content of lipid peroxides (TBA reactants) in the serum or in the cerebrospinal fluid.

3. CANCER

There is only one way to avoid making mistakes: do nothing or try nothing new.

Albert von Szent-György
(Nobel Prizewinner, 1937, for the discovery of vitamin C)

Cancer is a common disease: between every fourth and fifth individual contracts some form of cancer; in all of us, thousands, perhaps millions, of so-called cancerogenic changes take place, the antecedents of cancer. In most of us, however, the repair systems within the cells prevent the disease from becoming more advanced. If the defence system does not function the disease will break out. There is a growing consensus amongst experts that selenium, vitamin E, β-carotene, vitamin A and some essential fatty acids have cancer-preventive properties.

Cancer is the uncontrolled growth of cells to form a tumour which, in some cases, can invade the surrounding tissues of the body and spread, by the process of metastasis, to form secondary tumours in other parts of the body.

Although this description is true of most cancers, it should be noted that cancer is not one disease; there are many different forms of cancer. This is the case because the body is made of many different organs and many different tissues. Cancers arising in different organs are composed of different types of cancer cells (lung cells, skin cells or blood cells, for example) and they have different effects on the patient. They have different prognoses and may also require different forms of treatment.

The commonest forms of cancer in the UK are cancers of the lung, the breast and the skin. Of these, *lung cancer* is still the commonest and accounts for one out of every six cancer patients. The main cause of lung cancer is cigarette smoking. It is estimated that smoking lies behind 90% of all lung cancer cases. Moreover, lung cancer is very difficult to cure. Only one out of every ten patients survives for longer than ten years.

Another common form of cancer in Britain is *skin cancer*. Every year twenty-five thousand new cases of skin cancer are reported. It appears that some forms of ultraviolet light can cause skin cancer and that these cancers occur more frequently amongst people who do a great deal of sun-bathing (e.g. in Spain). However, most cases of skin cancer are relatively easy to treat and can often be cured completely, if discovered at an early stage.

The commonest form of cancer for women in Britain is *breast cancer* and affects about one woman in twenty at some time in her life. There are, however, good chances for a cure if discovered early enough by routine self-examination for lumps on or in the breasts. Unfortunately, too many women do not take the trouble to check until it is too late. As yet, the causes of breast cancer have not been discovered.

Cancer of the liver is relatively rare in the UK, where only a few hundred cases are recorded each year. On the other hand, this is one of the commonest forms of cancer in the underdeveloped world. One of the primary causes of liver cancer would appear to be the hepatitis B virus, which is especially rife in many Third World countries. There is no effective treatment for cancer of the liver and patients almost always die within a couple of years.

One out of every three cases of cancer, and perhaps even up to 40% of all cases, is estimated to be due to nutritional factors. This is partly because there are a great number of carcinogenic substances in foodstuffs, and partly owing to the fact that many of us are not getting enough cancer-protective micronutrients. Population studies implicate dietary fat intake in the etiology of colorectal (large bowel and rectum) cancer, and the incidence of colorectal malignancies around the world is positively correlated with meat and fat consumption and total calorie intake. When lipid molecules in our cells break down, it is usually by a chemical process called peroxidation. Evidence has accumulated over many years which suggests that lipid peroxidation may have some connection with cancer. There is no longer any doubt about the fact that certain micronutrients and fibre prevent cancer. This has been proven by about 30 major population studies carried out in Finland, Sweden, Holland, Switzerland, Britain, the United States and other places. To my knowledge, at least 24 intervention studies in which vitamins and minerals are administered as supplementation treatment to cancer patients are under way at the time of writing.

Cancer is usually treated surgically, with radiation treatment or with cytotoxic drugs (chemotherapy). Immune therapy, vitamins, minerals and essential fatty acids are generally viewed as supplementary treatments. The progression of the disease depends to a great extent on the treatment and on the patient's resistance. There are, to date, no studies on the treatment of cancer exclusively using micronutrients. Nevertheless, isolated cases have been reported where secondary tumours (metastases) have diminished or have totally disappeared in the course of a treatment with vitamins, minerals and essential fatty acids. It has also been established that selenium and vitamin E protect the liver and heart from the effects of a certain cytostatic, adriamycin.

The utilization of vitamins, minerals and essential fatty acids in supplementation treatment for cancer is based on the knowledge that these substances have a cancer-preventive effect, and on the fact that those people who contract cancer already had, in many cases, insufficiencies of these micronutrients when they were healthy. Operations, radiation treatment and chemotherapy do nothing to readjust the vitamin deficiency condition; on the contrary, radiation treatment and chemotherapy dramatically reduce the blood levels of β-carotene and vitamin E and the deficiency has to be counterbalanced with food supplements. Moreover, it is not just a question of the need to re-establish a balance. The micronutrient balance of cancer patients has to be optimal, so it is often necessary to administer doses which are larger than RDAs in order to maintain satisfactory levels.

Furthermore, it would appear that β-carotene and canthaxanthine (provitamins A), selenium, fish oils, and gamma-linolenic acid inhibit the proliferation of cancer cells. In addition, these micronutrients strengthen the individual's defence system and may thereby also protect against the development of metastases (see 1.6 on the immune system). Supplementation treatment bolsters up the general condition and assists the patient's capacity to deal with the strain imposed by radiation treatment and chemotherapy. In my own supplementation treatments of cancer patients I attempt to raise the selenium blood levels to 200–350 μg l^{-1}. (The normal levels for the average Finn are about 100 μg l^{-1}.) In my opinion cancer patients also need antioxidants like vitamins A, C, E, B1, B2, B3, B6 and some amino acids. In addition I recommend manganese, magnesium and the essential fatty acids γ-linolenic acid and fish oils, EPA and DHA to cancer patients.

Once in a while an argument is presented against vitamin supplementation of a cancer patient, on the grounds that vitamins might 'feed' the tumour and eventually give rise to metastases. This has never been scientifically documented, and hence must be viewed as pure speculation. Another argument is that the efficacy of supplementation therapy lacks the backing of double-blind evidence. This is true, but it does not prove the contrary. In my view, a cancer patient cannot afford to wait for studies, which, after his death eventually might confirm the benefits of nutritional therapy.

A more detailed description of carcinogenic factors can be found in Chapter 3, section 11.

4.   HEART AND CIRCULATION

One of the worst nutritional sins which we can commit is the over-consumption of animal fats and salt. Eat less salt or replace it with a healthier alternative like mineral salt, where potassium and magnesium replace a proportion of the sodium.

Free radicals precipitate the production of rancidified fats (lipid peroxides) in the blood and in the blood vessels. When the blood is deficient in antioxidants, the dangerous cholesterol-rich LDL is oxidized, causing arteriosclerosis (hardening of the arteries). Free radicals are also the culprits in a new concept in the field of

cardiovascular diseases, reperfusion injury. Antioxidants, magnesium and essential fatty acids protect the heart and the blood vessels from injury.

It is a great pity that cardiologists and neurologists have not shown more interest in free radicals, reperfusion injury and antioxidants, despite the fact that these factors are of immense significance in circulatory diseases.

A Danish research group (which included the Nobel Prizewinner, **Henrik Dam**) discovered, as far back as the 1950s, that sclerotic (hardened) blood vessels contain large quantities of rancidified fats, i.e. the products of oxidation. It took no less than 35 years, before this observation could be confirmed in a recent study at King's College Hospital in London. It was shown that serum lipid peroxides are, in fact, higher than normal in patients suffering from myocardial infarction and arteriosclerosis. It is presumed to date, that free radicals oxidize the LDL-cholesterol (to o-LDL), which then forms hardening plaques in the artery walls. Oxidized substances narrow the blood vessels and inhibit the working of an important enzyme called prostacyclin synthetase, which catalyses the production of the vital substance prostacyclin (also known as PGI2). When the level of PGI2 drops in the vein walls, the blood gradually begins to clot at that spot. The red blood corpuscles are damaged as iron and copper flow out of them, which further stimulates the oxidation process and a vicious circle is set in motion which leads to the growth of the blood clot. Sooner or later it tears itself loose and travels around the circulatory system until it gets stuck in one of the tiny veins in either the lungs, the brain or the heart.

Certain natural dietary substances, vitamin B6, γ-linolenic acid, selenium and fish oils for example, may prevent the clotting of blood platelets (thrombocytes) and thereby inhibit thromboses.

Blood circulation is poor in constricted coronary arteries which provide the heart muscle with blood and oxygen; consequently the patient develops oxygen deficiency (angina pectoris) when he exerts himself. Oxygen deficiency and the subsequent resumption of blood supply (reperfusion) precipitate the production of toxic oxygen radicals in the heart cells. This increases the formation of oxidation products (lipid peroxides) and may lead to cell damage. Here we have another recently discovered phenomenon, called **reperfusion injury**.

Reperfusion injury

Reperfusion injury or oxidation damage occurs when cells, which have been subjected to oxygen deficiency due to reduced blood supply, are damaged, paradoxically enough, when the blood flow returns to normal and the tissues are again fed with oxygen (reperfusion). It has been discovered that reperfusion, because of a number of biochemical reactions, leads to the formation of large quantities of oxygen radicals which damage the cells. This phenomenon was first described by a Swedish professor, **Karl Arfors**, now at the University of California in San Diego. Arfors, and several other researchers, have also shown that when the tissues are protected by antioxidants the damage is reduced — even when this is first applied some time after the damage occurs. In the course of the last five years medical journals and scientific symposia have increasingly drawn attention to the phenomenon of reperfusion injury. Unfortunately this interest does not appear to be shared by many doctors who still undertake no measures to prevent the problem.

This form of oxidation damage is a risk factor in coronary artery disease, thrombosis, arteriosclerosis, brain circulatory disturbances and in all forms of surgery. This is especially the case in transplants, in hand and heart surgery, where the blood supply to the tissues is shut off for a long time, with reperfusion injury as a consequence. It is now a recognized fact that antioxidant treatment hinders the harmful effects of oxygen radicals in cases of arrhythmia, for example. Antioxidant treatment can also prevent heart failure if the treatment is given in time, i.e. within three hours after the heart attack.

We both can and ought to supply our tissues with adequate quantities of antioxidant protection against circulatory disorders. If we consume sufficient amounts of vitamin E, the heart will be better equipped to resist oxygen deficiencies and reperfusion. We still do not know exactly how much is needed to raise the heart's vitamin E content to the optimal level, but it is probably in the region of 400–800 mg/day. Professor **Kenth Ingold**'s laboratory in Canada is at present carrying out tests on this subject, so doubtless we will be hearing more about the field in the near future. In any case, it is already clear now that the present recommendation of 10 mg/day vitamin E is far too low in order to have any effect on oxygen deficiency or reperfusion injury. We also know that a combination of water- and fat-soluble antioxidants gives better protection against reperfusion damage.

Furthermore, there is a connection between reperfusion injury and arrhythmia: reperfusion injury increases the formation of oxygen radicals, which attack the fatty acids in the cell membranes and form toxic compounds (aldehydes and lipoxins), which can stimulate potentially lethal arrhythmia.

Antioxidants (the antidotes to rancidification) like β-carotene, ubiquinone, vitamin A, C, E and B6, together with selenium and zinc are able to inhibit these dangerous disturbances in the heart cells. Ubiquinone (also called coenzyme Q10) is a very important vitamin-like antioxidant substance which also supplies energy to the heart cells. Ubiquinone is commonly prescribed by Japanese doctors for heart patients. The normal dosage is 10 mg three times a day.

In experiments, both with animals and human beings, it has been shown that the heart can better tolerate oxygen deficiency (ischaemia) and reperfusion if the heart muscle contains enough antioxidants. Clinical experiences with synthetic and natural antioxidants confirm this beneficial effect, according to studies which have been published in both the East and the West.

Cardiologists have long known that low *potassium* causes arrhythmia. They now estimate that low *magnesium* is the culprit in one case in every 1000. According to my experience, ordinary blood tests (serum magnesium) usually only detect severe magnesium deficiency. Magnesium may quieten arrhythmia even when serum magnesium tests appear normal. Magnesium supplementation may not help everyone with arrhythmia but it is worth trying. Particularly patients on digitalis should be monitored for magnesium deficiency. Digitalis medication may lead to toxicity and arrhythmia in persons with low potassium and magnesium.

Danish doctors have recently found out that magnesium supplementation reduces reoccurrence of myocardial infarction and sudden death, by preventing fatal arrhythmia.

My advice to heart patients is to take 300–600 mg of magnesium daily: it will probably do some good for your heart and there is no danger of toxicity.

Finnish studies on cardiovascular diseases and antioxidants

In 1980, Dr **Kaarlo Jaakkola** and co-workers caused a doctrinal controversy when he reported that supplementation with selenium and vitamin E significantly alleviated symptoms of angina pectoris and improved the physical performance of ischaemic heart disease (IHD) patients. Based on experience in veterinary medicine the authors had administered a daily dose of nearly 2000 μg of selenium in the form of sodium selenate and 400 IU of vitamin E. Although the study lacked controls it stimulated public and scientific discussion, as well as further research in Finland and elsewhere, on the role of antioxidants and selenium in cardiovascular diseases. Subsequently, a Swedish study confirmed that 2 mg vitamin E and 40 μg of selenium per kg bodyweight every second day indeed prevented experimentally induced myocardial necrosis in pigs weighing between 70 and 90 kg.

So far, data from four prospective and one cross-sectional study has enabled the analysis of the role of selenium and selenium deficiencies in the causes of IHD in Finland. The prospective studies are based on follow-up groups examined from 1972 to 1977. In all of these studies the initial examination included information on cardiovascular history, risk factors and physical measurements. Stored blood samples were available on serum selenium, and in two studies also for serum fatty acid analysis.

From the North Karelia survey a cohort (group) of 8113 subjects, who were initially free from IHD, was formed in the early 1970s and was followed up (in retrospect) over seven years, from 1972 to 1979. During that time a subsequently clinical or fatal myocardial infarction was diagnosed for 252 subjects and 131 subjects died of other cardiovascular complications. From the same cohort, controls were matched to each case according to age, sex, smoking habits, serum cholesterol, diastolic blood pressure and history of angina pectoris. The cases initially had a significantly lower serum selenium concentration than their referents (52 as opposed to 55 micrograms per litre). An extremely low serum selenium level of less than 35 μg l^{-1} was associated with a 6.9-fold risk of coronary death, compared to those with a level of 45 μg l^{-1} or more. Serum selenium of 35–44 μg l^{-1} was associated with a 2.2-fold risk of myocardial infarction. The respective relative risks for other cardiovascular diseases were 6.2 for serum selenium levels of 35 μg l^{-1} and 1.5 at 35–44 μg l^{-1}. The study thus yielded a clear inverse association between serum selenium and fatal and non-fatal cardiovascular diseases. The lower the initial serum selenium (from the initial blood samples) the more serious was the outcome.

Another five-year follow-up of 12 155 initially healthy persons, both men and women, between the ages of 30 and 64 years, was designed to start in 1977. During the follow-up, 92 persons died from myocardial infarction. Again, controls were matched sex, age, serum cholesterol, mean arterial pressure, tobacco consumption and history of cardiovascular diseases. Serum selenium was slightly but not significantly lower among cases than controls (62 versus 68 μg l^{-1}), but a serum selenium level of less than 45 μg l^{-1} was associated with a 3.2-fold risk of coronary death. Serum arachidonic acid concentrations correlated strongly with serum selenium concentrations among the controls and somewhat among the cases. Serum arachidonic acid levels also correlated with eicosapentaenoic acid (EPA) concentrations (a polyunsaturated fatty acid mainly found in fish) among the controls, probably reflecting the consumption of fish in the diet. The prime result of this study was that serum fatty acid content was associated with the risk of death from myocardial infarction in healthy persons with a low dietary intake of essential fatty acids.

A possible link between low serum selenium and reduced HDL-cholesterol (the 'good' cholesterol), with the resulting risk of coronary heart disease, has been suggested by stored samples studied by Dr Luoma and co-workers. During a five-year follow-up period, 141 men had died from cardiovascular diseases, 105 of them from myocardial infarction. In the total cohort, deaths from all causes and cardiovascular deaths were associated significantly in this study, where 26 healthy subjects were given a daily supplement of 96 μg of selenium and an improvement in the ratio of HDL- to total cholesterol was noted. The researchers hypothesized that selenium supplementation may, in subjects with low selenium status, reduce the risk of ischaemic heart disease. This theory was substantiated by a recent study of coronary risk factors among 1 132 Finnish men, aged 54 years. A serum selenium level of less than 85 μg l^{-1}, associated with low serum HDL-cholesterol, increased thrombocyte (blood platelet) aggregation and ischaemic exercise ECG (electrocardiogram) findings.

The notoriously high IHD mortality in Finland has decreased since the 1970s. The decrease has been, at least in part, attributed to an increased intake of selenium and other nutritional antioxidants.

Some case stories

My uncle, born in 1921, was one of the first heart patients whom I treated with antioxidants, in the early 1980s. He suffered from severe constriction of the coronary arteries and experienced debilitating chest pains at the slightest exertion. Opening the garage door on a cold winter morning had become almost an impossible task for him and he could hardly walk more than ten metres without pain. His condition deteriorated rapidly and it seemed that he would not be able to survive much longer. He was repeatedly driven to the local hospital by ambulance, where he was treated with nitroglycerin and beta-blocking agents, but to no avail. An exercise electrocardiogram (ECG) test showed that his coronary heart disease was undeniably very serious. This was in 1981.

That autumn, I prescribed for my uncle 900 μg of inorganic selenium (sodium selenate), 400 μg of vitamin E and 600 μg of vitamin B6 daily. In the space of a few weeks his symptoms were alleviated and he gradually began to be able to tolerate more physical exertion. The following winter, he was again able to enjoy his walks in the nearby forest without suffering any chest pains, even to the extent of walking through snow up to his waist. His neighbours were astonished at the remarkable effect which selenium and the vitamins had on him. In February 1982 he felt convinced that he was completely healthy again and wanted to stop the selenium and vitamin treatment. I forbade this, because I knew that otherwise his condition would deteriorate in the space of just three weeks. His healthy condition continued unchanged until 1983 when he stopped the treatment, in spite of my warnings. As expected, he became ill again but after the resumption of the treatment his condition again improved within a few weeks. At the time of writing (May 1989) his condition, at the age of 68, is still much better than it was in 1981. He attributes his health to antioxidants, which he still continues to take regularly.

Since then I have treated hundreds of heart patients for similar ailments. While I was working as the company doctor for IBM Finland, one of my patients, a man of about 50, complained that when he took his daily walk around the IBM building in his lunch-break, he had begun to get chest pains after he had covered a certain distance. An exercise ECG test showed severe signs of oxygen deficiency (ischaemia) in the heart muscles. The patient was given selenium, vitamins E and B6, plus nitroglycerin to be used when required. I asked him to come back in two months, by which time the pains had gone and he was again able to resume his lunchtime walks.

The effects of selenium, ubiquinone and the other antioxidants usually take about three weeks before they have an impact on chest pains, although some patients respond after a week while with others it may take as long as three months. In some instances the dosages may have to be increased before effects can be observed. Some patients have attempted to end the treatment on their own initiative but this has most frequently resulted in the reappearance of their symptoms.

A 60-year-old Swedish nurse came to me suffering from severe arrhythmia (abnormal heart rhythms). In Sweden the cardiologists could see no alternative to providing her with a pacemaker. Her pulse was so slow that she could no longer be bothered to walk up the stairs, let alone work at her demanding job. Neither myself nor any of my colleagues had any experience, at that time, in 1982, of treating such arrhythmia with antioxidants. Many of my heart patients who had been treated with selenium had reported an improvement in heart rhythms but this patient did not merely have occasional disturbances in the heart rhythms, she had a disturbance in the transmission of impulses between the auricle and the ventricles. I explained to her that I had no idea to what extent selenium therapy would help, but that the treatment was, in any case, quite safe. She agreed to give it a try and I prescribed selenium and vitamins.

Three months later I received a letter from her, in which she expressed her gratitude for the improvements in her condition and stated that her arrhythmia had disappeared. Somewhat later she came for a check-up and for a renewal of her prescription. A routine ECG was completely normal and years later her condition was still good.

Recently, a number of double-blind studies on antioxidants have appeared in several scientific journals. These have been concerned with the capacity of antioxidants to prevent reperfusion injury and to stabilize the membranes of the heart cells and the resultant inhibition of the incidence of arrhythmia. In addition, antioxidants seem to decrease the extent of damage to the tissues after experimentally induced coronary ischaemia in pigs, dogs and other experimental animals. It would appear that scientific studies are now confirming quite beautifully the experiences which I have had with my own patients from the early 1980s.

No treatment has a 100% record of success with all patients. This applies equally to supplementation treatment. Some of my patients with coronary artery disease are not totally relieved of chest pains, although most of them do feel fresher during the treatment. Some patients have reported an alleviation of their arrhythmia although this has not always been confirmed by their ECG tests. None of the patients have complained of serious side-effects from antioxidant and magnesium treatment.

Avoid salt

The big international Intersalt study has recently drawn the attention of all doctors to the part played by salt in cardiac and cardiovascular diseases. Our foodstuffs already contain so much salt that it is quite unnecessary for us to add more table salt! Avoid this and you can safeguard your heart.

5. ALLERGIES

Allergies are caused by a flaw in the immune defence, where both the cell immunity and the humoral immune system overreact to allergens in the environment or in food. Nutritional support with γ-linolenic acid and fish oils together with magnesium, selenium, calcium, zinc and vitamins can build up the immune defence against allergies.

The Finnish freelance journalist, **Reijo Ikävalko**, visited me in 1986 in order to interview me about my methods of nutritional treatment. Amongst other things, he asked me about which illnesses I had the best results with. I told him that I had investigated and treated successfully a great number of allergic patients. He then managed to convince me to try the method on him, because, as he explained, he had been plagued by allergies to pollen, dust, dogs and many other things for more than 25 years. His symptoms were a streaming nose, coughing and breathing difficulties. He later admitted, by the way, that his intention was to expose me in the press as a fraud if the treatment produced no results. He had no faith whatsoever in any kind of 'nutritional doctor'.

It did not turn out that way, however. Two months after the commencement of the treatment his symptoms had cleared up completely, using a combination of nine daily tablets of dolomite calcium, Bio-Selenium and Bio-Glandin (now also available in the UK). Like a good journalist, Mr Ikävalko asked the head of Helsinki University Hospital's allergy clinic about his opinions and experiences with allergies.

'Allergies come and go of their own accord,' said the allergologist, 'and you have been cured by yourself. This has nothing to do with vitamins. You can quite happily stop taking these superfluous pills now, because you get all the necessary vitamins from your daily diet. If your allergy was due to vitamin deficiency then it would reappear as soon as you stopped taking the vitamins, but I don't think that will happen!'

Alas for Ikävalko! He followed the advice and stopped the supplementation treatment and the allergy did indeed erupt again. Not surprisingly he elected to resume the treatment and the allergy symptoms disappeared again. He has since been able to pet dogs, which used to force him to run from the room because of 'hay fever' symptoms, and tells me that he would rather take a few pills every day than suffer his unpleasant symptoms. As a journalist, and in contrast to his previous views, he has recounted his experiences through the mass media and is probably the most renowned allergy case in Finland to be cured by natural methods.

Nine-year-old Lena came to my surgery one day with her mother. Lena was suffering from a milk allergy and asthma and she had given up horse-riding because of an allergy to horses. Treatment with γ-linolenic acid, zinc and selenium produced results within a week and all her symptoms disappeared. At the ensuing check-up she showed no signs of a rash or of respiratory problems. Now she can make do with two capsules of γ-linolenic acid daily and she can participate in games at school and other sporting activities, including riding, which previously had been impossible for her.

I have treated several hundred asthma patients with vitamins, minerals and essential fatty acids. Generally, they experience a noticeable reduction in their respiratory difficulties and in mucus formation in the lungs, often within as little as a few weeks. Some patients have even been able to dispense altogether with other medicinal treatments such as asthma pills and inhalers. A 28-year-old nurse, for example, was recently cured completely of asthma, and pollen and food allergy, which she had suffered from for 20 years.

As a consequence of a large number of positive experiences, I recently carried out a study on 20 allergic children who were members of the Helsinki Allergic Society. Almost all of them had zinc and selenium deficiencies and very low blood levels of the fish oils, EPA and DHA. This was quite simply because they did not eat fish. In the course of a 10-week treatment with antioxidants and essential fatty acids, 90% of the children experienced a clear improvement, half of them becoming free of their symptoms. The symptoms returned shortly after the end of the treatment period but with the aid of food supplementation it was again possible to alleviate them or make them disappear. The study attracted a great deal of attention also in Sweden, where it was published in the book, Barnallergi (*Childhood Allergy*), by Margareta Calmgård-Bergmark (Allerbok, Stockholm, 1986). Recently, a few double-blind studies have also been published elsewhere on the effect of GLA and fish oil supplements in atopic eczema. In my experience, a combination of fatty acids, vitamins and minerals yields a better prognosis.

Unfortunately, the treatment is not effective for all asthmatics, especially for those who have used cortisone tablets regularly or over an extended period. Cortisone counteracts γ-linolenic acid and prevents the same degree of widening of the respiratory passages, prevention of mucous formation and inhibition of infections, as in those patients who have not used cortisone.

How do we explain the clinical effects of supplementation treatment in the case of allergies? Recently it has been discovered that *inflammation* in the bronchi plays an important part in the etiology and development of bronchial asthma. We now know that *free radicals* are involved in inflammatory processes. Asthma and other allergic responses also are stimulated, at least partially, via the prostaglandin metabolism and the immune system (see 1.6). People with allergies have higher blood levels of immunoglobulin E (IgE). The B-cells treat this high-molecular protein, apparently under the direction of the T-cells. As the person grows older the immune activity of the T-cells is weakened more rapidly than that of the B-cells. This is particularly noticeable in the case of the suppressor T-cells, whose task is to defend the body from reactions in the immune system which could be aimed at attacking the body's own tissues. Failing T-cell activity is connected with arthritis, asthma, other allergies and the so-called auto-immune disease.

It would appear that the weakening activity of the T-cells leads to intensified symptoms and repeated outbreaks of allergies. As far as I can see, we ought to try all forms of treatment which reinforce the workings of the T-cells, provided the treatment is safe. Zinc, for example, has a stimulating effect on the activity of the hormone, thymulin, and on interleukin-2. Fatty acids regulate the prostaglandin, leukotrien and thromboxane metabolisms and thus suppress the inflammatory responses. Antioxidants combat the free radicals which are generated by inflammation and theophyllin, a widely used asthma-medicine. In my opinion, the asthmatic therefore needs a daily supplementation of dietary antioxidants.

6.   SKIN DISEASES

Skin diseases are something which the doctor likes: the patient doesn't die but he doesn't get well either.

H. L. Mencken (1880–1956)

The skin becomes loose, wrinkled and ages more quickly when so-called cross-linkages occur between the proteins in the connective tissues (collagen and elastin). The skin loses its elasticity. Everyone can check the elasticity of their skin by pinching the back of the hand. Older peoples' skin takes longer to return to normal than that of young people. Exactly the same changes take place, also caused by chemical cross-linking, in the walls of the blood vessels: the vein walls harden. These changes in the condition of

**the skin are caused by oxygen free radicals, lipid peroxides and other harmful
substances.**

Of all the organs in the human body, the spinal cord and the brain contain most
fat. Next in fat content is the skin. The high fat content of the skin makes it especially
susceptible to attack by free radicals. People who expose themselves to alcohol,
tobacco and excessive sunlight have dry skin, which wrinkles more easily than other's
because these three factors promote free radicals and consequently the skin ages
more rapidly.

There are several means to avoid this happening. One possibility is to avoid too
much sun or to protect the skin with *para-aminobenzoic acid* (PABA). This is a
vitamin-like compound which protects the skin against the harmful effects of the
sun's rays. It is a component of many sun-tan lotions but can also be purchased
separately in chemist shops. Another natural sun-protector is *Aloe vera*. Antioxidants which are taken orally also protect the cells from damage by the free radicals
(see 1.5). The appropriate antioxidants are the micronutrients, ubiquinone, selenium and zinc, plus vitamins A, B6, C and E. There are also some fatty acids (EPA
and GLA) and amino acids which inhibit the formation of cross-linkages.

In the previous section I have described how allergic rashes can be treated with
vitamins, minerals and essential fatty acids. Rumours about the results of my
treatment spread and people with all kinds of different skin diseases began to appear
at my surgery. These included *acne, rosacea, psoriasis, scleroderma* (a disease which
causes the skin to become hard and stiff), *lupus erythematosus* (LED) and many
other diseases, some of which I had seldom, or never, seen before. There are
different forms of psoriasis — in some forms the skin flakes off at the elbow, in others
on the scalp; while some people get suppurating sores on the soles of their feet and on
the palms of their hands (*psoriasis pustulosa*).

In 1982, Professor **Lennart Juhlin** of Uppsala University in Sweden, showed that
certain skin diseases are connected with insufficient activity of the selenium enzyme,
glutathione peroxidase. Selenium and vitamin E helped some of the patients but the
results were not totally convincing. Dr **Ingrid Emerit** in Paris has, however, proved
that antioxidant treatment (in the form of SOD injections) has some effect against
scleroderma.

One of my psoriasis patients, a young man with *psoriasis pustulosa*, had been
treated at the university hospital for several years but without success. He lived more
than 300 miles from my clinic and in the beginning I had to disappoint him by
explaining that I had never treated anyone for this rare disease before and that I had
no idea whether or not vitamin and mineral treatment could help him. Nevertheless,
he was willing to try and a few months of treatment were not going to be very
expensive. In 1983 I prescribed selenium, zinc and vitamins A, B, C and E to be
taken over a period of three months. When the three months were over and he came
back for a check-up, his hands and feet were almost back to normal. He is still healthy
today but has to take his daily food supplementation, otherwise the rash breaks out
again on his hands and feet.

Of all the zinc in the body, 20% is concentrated in the skin which therefore reacts
quickly to zinc deficiency. When *acne*, for example, is treated with zinc the results
have been mixed. It would appear that the benefit is greater when zinc is combined
with vitamins, selenium and possibly magnesium and essential fatty acids.

A young woman who worked in a big hospital came to see me. She had let her hair grow very long because of her acne, and she always walked with her head bowed forward so that her face would be partially concealed by her hair. She had suffered from acne for 15 years and none of the dermatologists whom she had visited had been able to help her. We agreed to try a treatment with selenium, zinc and vitamins A, B, C and E plus some other minerals over a three-month period.

At the check-up she was happy and full of smiles. Her hair was still long but was combed back and she held her head up normally. Her skin was almost totally free of acne. I have similar experiences with other acne patients but the treatment has not been uniformly effective in all cases.

Rosacea is a rash which appears on the face accompanied by dark red patches and pus-filled spots. There is no medicinal treatment although cortisone sometimes helps. A young woman who suffered from this disease came to my surgery. She had been treated with cortisone for several years and now she wanted to try something safer.

The zinc and selenium levels in her blood were very low and I prescribed zinc, selenium and vitamins. One of her cheeks looked better after a few weeks but progress was slower with the other one. During the course of the treatment her zinc blood levels increased slowly but never reached up to normal levels. The patient has continued the treatment over a long period and although she has not been completely cured her condition has improved significantly — certainly more than it did under cortisone treatment.

Systemic lupus erythematosus (*SLE*) and *scleroderma* manifest themselves as chronic skin diseases. Both illnesses have a very varying development and when the diseases are active, cortisone tablets are the traditional treatment. Usually no treatment is given during less active periods. In the case of scleroderma, the skin becomes horny, while SLE patients get red patches, a 'butterfly' configuration, mainly on the cheeks but also in other parts of the body. Vitamin, mineral and fatty acid therapy have had a quite definite beneficial effect on these diseases in several of my patients. In some cases I have seen dramatic improvements within six months in diseases which had previously been untreatable.

My own daughter, **Kirsikka**, aged 20, suffers from chronic *atopic eczema* and for years dermatologists treated her with cortisone ointments. The ointments had a good effect but did not cure the illness. In the early 1980s I tried to treat her with zinc and other minerals and initially this seemed to help, but it was first after she received γ-linolenic acid (GLA) that a permanent improvement occurred. Today she takes capsules containing GLA, the fish oils, EPA and DHA, and Bio-Selenium, and if she forgets to take her medicine for just a few days, the symptoms return immediately.

As I mentioned in the context of allergies, I have treated many children and adults with allergic eczema who have been cured within weeks with essential fatty acids, selenium and zinc, β-carotene, and vitamins B, C and E.

If skin rashes are associated with some form of nervous complaint (neurodermatitis) then vitamin and mineral treatment can often end in disappointment. Sometimes, if the patient is suffering from a high level of stress, then even large doses will be unable to alleviate the body's irritation state.

Boils appear to be 'afraid' of vitamin E and selenium. A number of my patients who are susceptible to boils can be relieved of this problem with food supplementations of these two micronutrients. If they drop the treatment, the boils reappear within time.

It was established in the 1960s that β-carotene improves the skin's resistance to the harmful effects of sunlight. Recently doctors have begun to prescribe a combination of β-carotene and canthaxanthine as an effective means of combatting *hypersensitivity to sunlight*. Both of these antioxidants counteract and neutralize the free radicals which are caused by ultraviolet light and give the skin a protective gold-brown colour.

Dr **J. Fuchs**, of the Goethe University in Frankfurt, West Germany, has recently shown how ultraviolet light reduces the quantities of antioxidants in the skin cells. Vitamin E content falls to 80%, ubiquinol-9 to 38% and ubiquinone-9 to 57% of the normal level. These studies provide us with an explanation for why sunlight is damaging to the skin: it precipitates free radicals which react with, and use up, the skin's normal antioxidant defences. When the antioxidant levels drop, the skin becomes defenceless against the oxygen free radicals which damage the skin and can even lead to malignant melanomas (skin cancer).

On the background of this new information, it would be wise to take antioxidant supplements when we expose the skin to excessive sunlight, whether this occurs on the beach, in the Alps or in the solarium. A suitable daily supplementation would consist of about 20 mg of β-carotene, 200–400 mg of vitamin E, 10–30 mg of ubiquinone, 15 mg of zinc and 100 μg of selenium taken when we sunbathe or travel to the south, to the Alps or to hot climates where sunlight is much more intense than we are used to. My patients often tell me that 30 mg of β-carotene alone helps with their skin problems in the sun. In addition, there are several testimonies from sun-allergic persons that also γ-linolenic acid appears to be protective against the harmful effects of sunlight.

7.   RECURRING INFECTIONS

Disease is infectious.

William Shakespeare (1564—1616)

Virus infection involves, at least in part, other mechanisms than a direct effect of virus multiplication. Recent findings indicate that during viral infection, oxygen free radicals are generated by cells of the immune system which are mobilized to fight the virus and by a substance called xanthine oxidase, which is present in excess amounts in blood and extracellular fluids.

Contrary to the previous view, the viruses themselves do not lead to the disease but to the overreaction of the immune responses of the host. Furthermore, it has been shown in mice that influenza can be prevented by administration of a free radical scavenger, in other words an antioxidant. The antioxidant injected into influenza-infected mice protected them from lethal influenza injections. Antioxidants apparently have broad preventive and therapeutic applications in infectious diseases, since free radicals are not exclusive to influenza, but contribute to pathology of a large number of other diseases. (Patients would often tell me, while we were discussing the

details of a treatment for their chronic complaints, other than infections, that they were no longer affected by infections which otherwise struck them frequently.)

These astonishing results were published in the well-recognized journal *Science*, 26 May, 1989. These findings nicely explain my clinical observations throughout the years that I have been supplementing my patients with antioxidants.

According to the latest research, vitamins, minerals and fatty acids help to strengthen the immune system (see 1.6). This applies to iron, selenium, zinc, magnesium, copper, vitamins A, C, D, E and B6 plus the fish oils DHA and EPA and γ-linolenic acid (GLA). It must, however, be emphasized that in problematic cases, involving several recurring infections per annum, the immune effect can only be achieved with much larger quantities than we can take in via daily diet. This necessarily involves that we consume these substances as food supplements in tablet or capsule form.

A striking development in children's micronutrient treatment is the reduction in or even absence of many of the commonest infectious diseases amongst children who have received the natural preparations which protect the immune system. This also applies to children who were only 6 months to 4 years old, so there can hardly be any question of the placebo effect, when these children have managed to avoid colds, ear infections, sinusitis and other infections.

8. ULCERATIVE COLITIS

Digestion exists for health, and health exists for life.

G. K. Chesterton (1874–1936)

Ulcerative colitis is a chronic, nonspecific, inflammatory and ulcerative disease of the large bowel, most often characterized by diarrhoea with bleeding. The ultimate cause of this serious illness is still unknown, but the failure is probably in the immune system.

During my years of practice with nutritional therapies, I have principally prescribed a 'battery' of nutritional supplements for patients suffering from ulcerative colitis and the response has mostly been positive. After a few weeks of supplementation the patients have usually reported that their bowel functions were quite normal and that the attacks of diarrhoea had stopped.

The nutritional therapy usually consists of the following: 100 µg of organic selenium (L-seleno-methionine), 15 mg zinc (as gluconate), 10 mg β-carotene, 5000 IU vitamin A, one tablet of vitamin B complex, 200–400 mg vitamin E, 100 mg vitamin C, 300 mg calcium, 200 mg magnesium, 2–3 capsules of GLA from Borage Oil and 2–3 capsules of fish oils (EPA + DHA). With combination preparations, the daily amount of pills can be kept at a reasonable level.

9. DISEASES OF THE NERVOUS SYSTEM

The nerves — well, that is the whole person.

French proverb

It has become the accepted practice to use B vitamins as adjunct therapy for

disturbances in the nervous system and for mental stress. However, in this section I want to concentrate on diseases which have not previously been treated with vitamins, minerals and fatty or amino acids. Positive results have been obtained and there is every reason to initiate studies on the effects of these micronutrients for a wide range of neurological diseases.

Reperfusion injury in the brain

In November 1988, researchers into free radicals from all over the world met in Vienna to discuss new discoveries, including reperfusion injury. Researchers presented a series of studies which showed that circulatory disturbances in the brain cause cell damage, known as *reperfusion injury* (just like in the heart and the intestinal tract). This happens when blood flow returns to an artery which has been closed off, for example, because of a blood clot; large quantities of oxygen free radicals are produced during reperfusion, i.e. the restoration of the blood supply, and these react with the polyunsaturated fatty acids in the cells, forming cell-toxic compounds such as aldehydes and lipoxins. These substances damage the brain cells unless they are neutralized by an effective antioxidant defence, incorporating, amongst other things, vitamin C and vitamin E. It is therefore of the utmost importance, in the event of brain circulatory disturbances, to take supplements of antioxidants which can protect the most important cells in the body. These cells cannot be renewed!

Other neurological problems

A few years ago I read in *The Lancet* about a rather curious case history, published by my friend, the neurologist **Juhani Juntunen**: his patient, a young female lawyer, had been cured of an illness of the peripheral nervous system (polyneuropathy) with alcohol (*sic*). All other forms of treatment had been ineffectual. The result was paradoxical because alcohol usually causes disturbances in the peripheral nervous system.

A few years later I was visited by this very woman, who at that time was in such a bad condition that not even alcohol gave her any relief. I would by no means describe her as an alcoholic. She now wanted to try vitamin and mineral treatment, because, with the passage of time, she had become incapable of looking after her job due to numbness and pains in her legs.

After only a few weeks her condition, which in the two months prior to the treatment had been deteriorating constantly, began to show an improvement and two months later her symptoms had disappeared completely. Two years later she experienced a new attack of the disease and was again helped with vitamin and mineral treatment.

One day I was called up by one of my colleagues who also uses vitamins and minerals in his practice. He had witnessed an astonishing result in the treatment of a young woman who was suffering from *myasthenia gravis* (*MG*), a serious weakening of the musculature. The woman had recovered the use of most of her musculature during treatment with sodium selenate, an inorganic salt of selenium. This took effect so quickly that she could take a couple of capsules of this substance immediately before a strenuous piece of work and be able to withstand it. If she did not receive selenium the strength in her muscles rapidly faded.

This property of selenium had never previously been observed, so I suggested to my colleague that we should have the case thoroughly investigated by a neurologist and a neurophysiologist during a period in which the patient was hospitalized at the Institute of Occupational Health in Helsinki, where I was employed at that time. All the tests and examinations confirmed the validity of the patient's own experiences: after the ingestion of a couple of capsules of selenium the patient was able to lift her head and arms several times from a horizontal position while measurements of the electrical conductivity of the nerves and muscles showed almost normal values. Without selenium she was unable to lift her head or arms and this was confirmed by the accompanying neurophysiological measurements. This was a new and hitherto unknown effect of selenium, which apparently has an influence on the co-operation between nerves and muscles. In my view, a carefully controlled clinical trial ought to have been made, in order to research this phenomenon further, but unfortunately the University Clinic in Helsinki was not interested. On the contrary, they reported me to the National Board of Health for giving selenium to an MG patient. The case was dismissed, however.

Parkinson's disease has been treated experimentally with antioxidants in the United States and Canada since 1979. It was discovered that daily doses of up to 3400 mg of vitamin E and 3000 mg of vitamin C can postpone the advancement of the disease for several years. The results have been published in neurological journals and at scientific conferences. Double-blind studies are now being carried out on this effect.

Professor **Jørgen Clausen** has published a major study on about 100 residents in a Danish nursing home, where the elderly at the home were given an antioxidant cocktail, i.e. vitamins A, B6, C, E and the trace elements zinc and selenium (Bio-Selenium) as well as the essential fatty acid, γ-linolenic acid (Bio-Glandin). Professor Clausen was able to show, using advanced methods of measurement, that the elderly residents experienced an improvement in the blood flow in the brain. At the same time, psychological tests indicated an improvement in their mental performances. In addition, the levels of an age pigment called lipofuscin in the red blood cells were reduced. A control group who were given placebo also participated in this double-blind study.

Patients suffering from *amyotrophic lateral sclerosis* (ALS), a relentlessly progressive neurological disorder, have been successfully treated in a placebo-controlled study with branched-chain amino acids, L-leucine, L-isoleucine and L-valine. During the one-year trial period, 11 patients on placebo showed a decline in their functions, consistent with the natural course of the disease. Those (11 patients) treated with amino acids showed significant benefits in terms of maintenance of muscle strength and continued ability to walk. This study was published in *The Lancet*, 7 May, 1988.

10. ARTHRITIS AND RELATED DISORDERS

Most of my patients who have trouble with their joints are either suffering from rheumatoid arthritis or from osteoarthrosis. In many of these cases it has been

**possible to delay, or even arrest for some time, the progression of the disease using
vitamins, minerals and essential fatty acids to enhance the functioning of the immune
system.**

**Fatty acids, B vitamins, zinc, copper and selenium have revealed themselves to be
helpful, as an adjunct therapy, in the alleviation of pains and stiffness in the joints. In
addition, a new preparation of gold and other promising new antioxidants are on the
way for the treatment of rheumatic complaints.**

Free radicals, lipid peroxidation and some of the harmful prostaglandins apparently play an important part in both the etiology as well as in the development of
various rheumatic ailments. Excess free radicals are able to destroy fatty acids,
cartilage and the lubricating fluid in the joints. The cell functions are disturbed and
harmful prostaglandins cause an inflammation response, setting a vicious circle in
motion.

In time the cell membranes are destroyed while strong enzymes, known as
hydrolases, are released, which attack and destroy the joint tissues. The free radicals
also attack the collagen, the connective tissue protein of which 30% of all the protein
in our bodies is composed. The lubricants of the joint surfaces are damaged, various
enzymes cease to function, polyunsaturated fatty acids are rancidified and the
genetic material in the cell, the DNA, is attacked.

When the quantities of free radicals in the tissues rise, the production of the
antioxidant enzyme, superoxide dismutase (SOD), also increases as a defence
mechanism of the body. This enzyme attempts to combat the free radicals and
thereby protect the cells from damage. Two different types of SOD enzyme are
found in the body: one needs copper and zinc (CuZnSOD) and the other one needs
manganese (MnSOD) in order to be able to function. There are large quantities of
SOD enzyme in the body; in fact it is the most commonly occurring protein in the
body after collagen, albumin, globulin and haemoglobin.

Attempts have been made on racehorses, and to a certain extent on human
beings, to inject SOD enzyme directly into the affected joint. Unfortunately the
effects are so short-term that its clinical use is still very limited.

Rheumatoid arthritis is a so-called auto-immune disease, because it is the body's
own immune system which attacks the joint-surfaces and the joint-lubricants.
Measurements of arthritic joints have revealed that they contain higher than normal
free radical activity, and a number of attempts have been made to treat them with
antioxidants which minimize the formation of free radicals and stabilize the functioning of the immune system. Not all of these experiments have enjoyed equal degrees
of success but this should not deter us from persevering with various combinations of
antioxidants, selenium, zinc, β-carotene, vitamin A, B, C, E, ubiquinone and
essential fatty acids.

Clinical practice, long used in Germany, has shown that copper and zinc, in
organic form, can alleviate pain and remove stiffness in the joints. Previously, some
German doctors used inorganic minerals but have given them up in favour of organic
minerals, partially because the inorganic types produced unwanted side-effects and
partially because the organic forms have been discovered to have a better
bio-availability.

The use of gold in the treatment of arthritic complaints was initiated by the
French doctor, **Jacques Forestier**, in the 1930s. At that time it was believed that

because gold was the most valuable of all the elements, it must also be beneficial in some way. Gold therapy became generally accepted without any basis in double-blind experiments to prove its effectiveness; today we know that gold is toxic for the body. Nevertheless, it is used with considerable success in the treatment of rheumatic and arthritic diseases, even though the mechanism of action is not yet known. Recent Japanese studies indicate however, that gold probably acts as an antioxidant. Thus other antioxidative trace elements, i.e. zinc, selenium and certain vitamins, might complement the clinical effect of gold.

Gold has hitherto been administered in the form of injections and the incidence of side-effects is fairly high. There are now, however, preparations of gold in tablet form which can be more easily tolerated.

The natural antioxidants are a completely non-toxic treatment which the patient can administer himself to a certain extent. The antioxidant tablet, incorporating selenium, zinc, vitamins A, B6, C and E, together with organic copper, γ-linolenic acid and fish oils, would constitute a good supplement for a rheumatoid patient.

A double-blind study in Scotland has shown that a combination of γ-linolenic acid and fish oils may alleviate arthritis symptoms and may reduce the patient's dependence on pain-killing drugs. Similarly, German and Austrian experiments and clinical practice reveal that a combination of anti-arthritis drugs and vitamins B1, B6 and B12 also have a pain-killing function in rheumatoid arthritis. This adjunct vitamin therapy is currently commonly given in university hospitals in Germany and Austria.

11. AGEING

The only way to live for a long time is to get older.
                                                    Francis Auber (1782–1871)

Everybody wants a long life but nobody wants to get older.
                                                    Jonathan Swift (1667–1745)

One of our most cherished wishes is to grow old slowly and gracefully. We would all like to be fresh, energetic and active pensioners and enjoy a long retirement. Why is it then that so many of us grow old before our time? What is it that ages and how can we avoid it? These questions have always preoccupied mankind but only now are we beginning to get closer to the secrets of old age. Free oxygen radicals appear to be an important factor in senile decay and the cell damage which is the source of most diseases. If this theory is correct, then it is possible for the individual to hinder senile decay to a great extent, and to slow down many of the visible signs of the body's decline. In this section I want to tell more about the reasons why we age prematurely and about which vitamins and minerals prevent illness, tissue death and senile decay. It has been shown recently, in nursing homes for the elderly in Finland, Denmark and Poland, that it is not too late to delay the ageing process.

The record for the oldest (properly documented) age in the world belongs to **Shigechiyo Izumi** from Japan, who lived to the ripe old age of 123 years. Fantastic stories of 160-year-old men in Georgia (Soviet Union) are not based on facts. It was common practice in Georgia for the son to take on the father's name in order to lie about his age and thereby avoid military service. This explains why it was only men who claimed to have reached such phenomenal ages, in contradistinction to other parts of the world, where mortality statistics are more reliable and where women live longest.

Average life-span has increased dramatically in the 20th century. Around the turn of the century a newly born baby boy could only count on living for 45 years, while for a woman the average was about 50 years. Today the average life-span is approaching 80 years but it is still women who live longest. The prolonged life-span is first and foremost due to the significant reduction in infant mortality, the eradication of many epidemic diseases and the improvements in health care. The maximum life-span of human beings has not changed for several centuries, however.

When does ageing begin? When we are 25 years old! This is no joke. Top athletes and sports stars are veterans at 25, and old when they are 30 years of age. It has been estimated that our physical power and performance decreases by 1% per year from when we are 25. For most of us this has little significance until we reach our 60s. Thereafter more and more of us begin to feel the hitherto unavoidable signs of *senescence*: the least exertion is enough to cause fatigue and shortness of breath; the skin loses its elasticity and our wrinkles become much more visible; we develop arthrosis and stiffness in the joints and our posture becomes more stooped; the hair turns grey and falls out; our teeth decay more rapidly, and so on.

If we look around us, we discover that not everybody ages at the same speed. There are many 70-year-olds who play 3–4 hours of golf or tennis every day, who swim goodness knows how many lengths at the swimming pool or who perform other physical feats which would be out of the question for many 45-year-olds.

At the beginning of our lifetime we are all programmed for a certain life-span which is determined, to a certain extent, by the genes. In the course of our lifetime we affect the 'biological clock' by our living habits, with the result that we either live longer or shorter lives than we are entitled to according to our genes.

Geriatricians are more or less agreed that it is the condition of the cells which determines whether we live long and stay fresh or whether we grow old before our time (for more about the life and death of cells see 1.4). Agreement is not nearly as clearly seen in the question of the extent to which we can affect the length of our own lifetime — apart from the well-known factors: don't smoke, don't drink too much alcohol, don't eat too many fatty foods and don't avoid exercise. Unfortunately, these habits have a synergistic effect, i.e. they reinforce each other and, sadly, they all have a tendency to appear in the same individual.

Our defence against disease is never stronger than the weakest point. This weakest point varies from individual to individual and we have all inherited tendencies to certain diseases, even though these diseases are not actually hereditary. The changes which occur in old age are due to our diminishing ability to defend ourselves against disease. But it is possible to fortify the defences.

Dr **Markku Halme** and I carried out a double-blind study on just these effects of vitamins and minerals on geriatric patients in 1983–1984 at a nursing home in the Finnish town of Hartola. Two identical groups of elderly residents, each containing 15 persons, were selected for the study which stretched over a year. In this period, the treatment group were given a supplementation of selenium and vitamin E, while the control group received a placebo medicine, which looked the same but which contained no active ingredients.

During the experimental period both groups were regularly checked and they were given blood tests. The study showed with great statistical significance that vitamin and mineral supplementation improved the participants' physical and psychological well-being: they were better able to deal with physically demanding tasks, they had a decreased tendency to depression and were more outgoing, while the patients in the placebo group did not experience comparably positive results.

The only error with the above study was that the number of participants was far too small. Three further double-blind studies were therefore initiated, one in Denmark, one in Finland and one in Poland, all in nursing homes for the elderly. Once again the control group were given a placebo medicine while the treatment groups in Denmark and Finland were given antioxidants (selenium, zinc and vitamins A, B6, C and E) for a one-year period. In addition they received the polyunsaturated fatty acid, γ-linolenic acid. In Poland, the elderly participants in the study received vitamin C and/or E, or the respective placebo.

The results show that the elderly participants had large quantities of lipid peroxides (rancidified fats) in the blood but that this could be counteracted with the antioxidant treatment. Likewise, the quantities of the age pigment (lipofuscin) in the red blood corpuscles were reduced while the blood circulation in the brain was improved. Arrhythmia also disappeared during the antioxidant treatment period. Psychological studies confirmed the physiological results in that the elderly Finns showed a tendency to improved well-being and greater liveliness.

The results have been published in many scientific journals and at symposia in Europe, the United States, India, Nepal, Japan and China. It seems that it pays to take antioxidants from about the age of 40 years and thus prevent, in good time, the biochemical changes associated with ageing.

A new population study in the United States has demonstrated that people with a high intake of β-carotene, vitamin C and vitamin E run reduced risks of developing cataracts compared with a control group which had a low antioxidant intake (Arch. Opthalmol. (1988) **106** 337–340).

13. OSTEOPOROSIS

Osteoporosis (brittle bones) are so common amongst the elderly that we more or less consider them to be a natural phenomenon. Almost all woman who have passed the menopause suffer from brittle bones.

Osteoporosis means demineralization of the bones. Between the ages of 40 and 90 large amounts of calcium, magnesium, phosphorus, silicon, zinc and other minerals are lost from the bones. Brittleness in women's bones is hastened by falling oestrogen production and lack of exercise. There is no known effective treatment, despite attempts with calcium supplements, vitamin D, fluorine and oestrogen. This means that we must concentrate on prevention of osteoporosis, and the preventive treatment ought to be initiated long before the menopause.

Experts have proposed a considerable increase in the recommended daily allowance for *calcium* for women over 35 years of age, from 800 mg to 1500 mg. This alone, however, is not enough, because in order to consolidate itself in the bones calcium needs to be accompanied by *magnesium*. Calcium supplements alone are apparently inefficient in preventing osteoporosis.

LITERATURE

Bittiner, S., Cartwright, I. and Tucker, W. *et al.* A double-blinds, randomized placebo-controled trial of fish oils in psoriasis. *Lancet* **1582** (1988) 378–380.

Bjørneboe, A., Søyland, E. and Bjørneboe, G., *et al.* Effect of dietary supplementation with eicosapentaeonic acid in the treatment of atopic dermatitis. *Br J. Dermatol* **117**, 5 (1987) 463–469.

Clausen, J., Achim Nielsen, S. and Kristensen, M. Biochemical and clincial effects of an antioxidative supplementation of geriatric patients. A double blind study. *Biol. Trace Elem. Res.* **20** (1989) 135–151.

Clausen, J., Egeskov Jensen, G. and Nielsen, S. A. Selenium in chronic neurological diseases: Multiple sclerosis and Batten's disease. *Biol. Trace Elem. Res.* **15** (1988) 179–203.

Dworkin, R. Linoleic acid and multiple sclerosis. *Lancet* (1981) 1153–4.

Elevated dosages of vitamins. Benefits and hazards. International Symposium, Interlaken, Switzerland, 7–9 September 1987, Abstracts.

Friedrich, W. *Vitamins*. Walter de Gruyter. Berlin–New York, 1988, p. 1058.

Gerbershagen, H. U. and Zimmermann, M. (eds). *B vitamins in pain*. pmi Verlag, GmbH, Frankfurt, West Germany, 1988.

Gey, F., Brubacher, G. B. and Stahelin, H. B. Plasma levels of antioxidant vitamins in relation to ischemic heart disease and cancer. *Am. J. Clin. Nutr.* **45** (1987) 1368–1377.

Macarthy, M. S., Rationales for micronutrient supplementation in diabetes. *Med Hypotheses* **13** (1984), 139–151.

Oda, T., Akaike, T. and Hamamoto, T. *et al.* Oxygen radicals in influenza-induced pathogenesis and treatment with Pyran Polymer-Conjugated SOD. *Science* **244** (1989) 974–976.

Tolonen, M. Vitamine, Mineralien und essentielle Fettsauren ais ergänzende Biotherapie bei rheumatischen Erkrankungen – erste Erfahrungen. In: Schmidt, K. and Bayer, W. (eds) Mineralstoffwechsel und rheumatischer Formenkreis. Verlagfür Medizin Dr Ewald Fischer, Heidelberg 1986, 93–97.

Tolonen, M. Finnish studies on antioxidants with special reference to cancer, cardiovascular diseases and ageing. *Int. Clin. Nutr. Rev.* **9** 2 (1989) 68–75.

3

Diet

1. IS OUR DIET HEALTHY?

Eat to live, don't live to eat.

Finnish proverb

Your food is your medicine.

Hippocrates (460–370 BC)

In the course of the last 20–30 years the food industry and the range of foodstuffs have undergone distinct changes. We now have a much wider choice of things to eat but we consume far more preserved and 'ready-to-eat' meals. In the past, transport and storage problems limited the available choice but these difficulties have been solved by the use of preservatives, additives and deep-freezing. Consequently, the 'shelf-life' of many foodstuffs has been greatly extended.

The incidence of food allergies has increased, in step with the growing use of additives and preservatives, to the extent that one person in five suffers from some form of allergy.

A great proportion of the vitamins and minerals are lost during the process of production and this is in part compensated for by the addition of artificially manufactured vitamins. Processed foods have become more refined and easily digestible, with the result that their fibre content has dropped considerably. This has resulted in a great increase in the incidence of 'diseases of civilization', such as constipation, stomach ulcers and intestinal cancers. The changes which have taken place in our dietary habits have not all been negative, however; there have also been positive trends in recent years.

The food consumption of more health-conscious individuals has changed in the direction of decreased intake of fats, meat, eggs and sugar while consumption of whole-wheat products, fruit and vegetables is gradually increasing. This is undoubtedly a beneficial development. On the other hand it is less positive, to say the least, that:

• the mineral (e.g. magnesium) content of foods is decreasing
• we use too many additives

- every third case of cancer is due to carcinogenic substances in foodstuffs and to the lack of anti-carcinogenic vitamins, trace elements and fibre in the diet
- the incidence of food allergies is on the increase.

Unfortunately we tend to pass our unhealthy dietary habits on to the next generation. The table below outlines roughly what sort of changes in our eating habits can have a beneficial effect.

Eat less	Eat more
Fat	Whole-wheat products
Meat	Fish (30/gr day)
Sugar and foods containing sugar	Potatoes
Snacks containing fat and sugar and 'between-meal' snacks	Vegetables and root vegetables
	Fruit and berries
Salt	

Only 30% of our total energy requirement should consist of fats and of this only a third, at the most, should be animal fats. Sugar consumption should not exceed 10 kg per person per year or 30 g per day (the average annual sugar consumption is now 70 kg). Daily fibre intake should be 30–40 g as opposed to the present average of less than 20 g. Salt consumption should not exceed 3 g per day. The average consumption of meat and milk products would appear to be at about the right level, although it is advisable to eat more fish instead of meat.

The debate on the relative advantages of biodynamically grown vegetables compared to those cultivated with the aid of artificial fertilizers and chemicals continues. The difference is perhaps not so significant, as some participants in the debate have maintained, because organic crops are exposed to the same degree of air pollution. On the other hand it is clear that the use of pesticides and chemical sprays should be kept at a minimum, and stringent controls should be enacted on a regular basis.

Food additives
The purpose of additives is to improve the storage-life, quality, taste, consistency, aroma and the colour of the product. At the present time over 5000 different additives are used in foodstuffs. Salt and spices were the earliest form of food additive but nowadays the use of vitamins and minerals has become increasingly common because so many of the essential micronutrients are lost in the course of food processing. Antioxidants (usually vitamins E and C) are employed in order to decrease the perishability and the risk of rancidification of foodstuffs. Antioxidants also prevent products from losing their colour. When an apple, banana or pear turns brown, it is because of oxidation on contact with the oxygen in the air. At this point the vitamin C is also oxidized to a pro-oxidant which can lead to cell damage. This is the same process as that which occurs in the cell membranes, which also derive benefit from antioxidants.

Nitrites are often added to foodstuffs in order to maintain a pleasing red colour and to prevent the proliferation of bacteria (in sausages and tinned meats, for example). Nitrites also occur in connection with frying or grilling of meat. Many vegetables have a high content of *nitrate* due to artificial fertilizers or air pollution; when these nitrates come into contact with the saliva in our mouths, they are converted into nitrites. When they reach the small intestine, nitrites are further transformed into nitrosamines which can be carcinogenic. Vitamin C counteracts the harmful effects of nitrosamines and thereby inhibits the development of stomach cancer. We can therefore reduce the harmful effects of nitrites by ensuring that we have a high intake of antioxidants.

Fibre is beneficial because it prevents the food from remaining in the digestive system for too long, thus limiting the length of time in which harmful substances (e.g. carcinogens) can cause damage to the system.

There is a great deal of controversy about the necessity of additives. The food industry maintain that they are essential while increasingly large numbers of consumers feel that they should be removed altogether from our food. The levels of additives should certainly be kept to a minimum because of the risk of food allergies, but scare campaigns have sometimes been exaggerated. The carcinogenic properties of food additives are dependent upon the degree to which such additives are consumed: we have the choice between normal consumption or gross over-consumption of these substances.

A few years ago there was a great deal of concern about the results of animal experiments which revealed the carcinogenic properties of cyclamate, an artificial sweetener. This substance had caused cancer of the bladder in rats, but it was shown that the rats had to take 50 times the permitted maximum dose for human beings before they contracted the disease. Of course the fact that a substance is carcinogenic at all is a cause for concern; but there is a risk involved in the misuse of almost any specific substance.

We *could* certainly survive without additives, but this would imply a smaller range of products, an increased perishability and higher prices for foodstuffs. If, for example, we stopped using nitrites in meats altogether, this would increase the risk of botulism, a fatal form of food poisoning. Botulism is very rare today, usually only occurring in fish which have been polluted or smoked privately, or which are eaten raw or marinated.

We are faced with the need to make a compromise: on the one hand we have to learn to live with food additives; on the other we must demand close controls over the quality and quantities of these additives.

To be on the safe side, it is advisable to increase our intake of β-carotene, ubiquinone, vitamins A, C and E, selenium and other micronutrients which protect us against the dangers represented by food additives.

2.   THE DIGESTION

To eat is human, to digest divine.

Charles T. Copeland (1860–1952)

For me, happiness is a question of digestion.

Lin Yutang (1895–1976)

Digestion begins in the mouth, from where the food is led down the oesophagus into the

stomach and, via the pylorus and the duodenum, to the small intestine. Here the food is mixed with enzymes from the pancreas and the liver via the gall bladder and the bile ducts. These enzymes break the food down to an even greater degree than has already happened in the stomach, enabling all the nutrients to be absorbed in the blood. The food which is not absorbed becomes waste which passes through the 12–30 feet of the small intestine and out into the large intestine before leaving the body.

The most important functions of digestion are the following:

• Carbohydrates (sugar) already start breaking down in the mouth.
• The food remains in the stomach for between 1–5 hours. The higher the fat content, the longer the food remains in the stomach. Here the breakdown of proteins begins under the influence of the gastric juices which contain hydrochloric acid and enzymes.
• Fats are broken down in the small intestine by enzymes from the pancreas and in the bile. Water-soluble vitamins are absorbed together with fats.
• Proteins, carbohydrates and vitamins are also absorbed in the small intestine. They are transported through the walls of the intestinal mucous membranes and absorbed in the blood stream. The nutrients are carried to the liver, which is the centre for food transformation, from where they are transported to those parts of the body where they are needed. If some disturbance in the system causes malabsorption, the body will only benefit from a limited proportion of the nutrients in the food.
• The bacteria in the large intestine break down the unwanted part of the food and a waste mass is formed which absorbs water — the result is the excreta which the large intestine pushes towards the rectum.

3.   HEALTHY EATING HABITS

I am convinced that the great secret of life lies in the digestion.

Sidney Smith (1771—1845)

Our dietary habits not only vary from individual to individual, but also from culture to culture. Doctors who adopt a more biologic approach to nutritional questions usually give the following guidelines for dietary practices which are healthy and which have a preventive effect on diseases:

• Try to eat each foodstuff in the most natural form possible. Avoid foods containing colour and preservatives and foods which have been sprayed with chemicals.
• Eat as little fat as possible — we consume enough fat in our daily diet whether or

not it can be seen with the naked eye. Note that vegetables, lean meats and fish contain essential fatty acids.

- Avoid sugar and foodstuffs which contain sugar, such as sweets, chocolate, soft drinks and jam.
- Eat whole-wheat bread and avoid white bread.
- Eat fresh vegetables, root vegetables and fruit and avoid tinned fruit and vegetables.
- Chew food thoroughly and eat in peace and quiet.

4.   PREGNANT AND NURSING MOTHERS NEED MORE

And the mother understands — as she always does in the end — instinctively.

Mika Waltari (1908–1976)

Vitamin and mineral requirements increase by between 10–50% during pregnancy and in the following period of nursing. This is not only because the foetus has its own needs, but also because the mother's oestrogen production leads to an increased formation of free radicals. Sufficient supplies of micronutrients ensure the normal development of the foetus and satisfy the baby's needs during nursing. Pregnant and nursing mothers should not only take supplements of vitamins and iron, as hitherto recommended, but also of zinc and selenium.

The pregnant mother does not need to eat for two, even though the vitamin and mineral requirement increases during pregnancy. It is the growing child which needs these extra micronutrients from the very moment of conception. In fact, from the baby's point of view it is in this early period that it is most important that its needs are fulfilled.

Towards the end of pregnancy, when the baby's skeleton, internal organs, musculature and other tissues are growing rapidly, zinc and selenium requirements increase.

During pregnancy the baby's vitamin and mineral needs are covered by the micronutrients in the mother's bloodstream. If she does not ensure a constant and adequate intake of the most important substances, it can lead to a deficiency condition for her, because the baby taps her resources. It is only later, if the levels fall even further, that this condition affects the baby.

The latest research suggests that the mother's vitamin and mineral status around the period of conception also plays an important role. According to a recent American study the risk of neural tube defects (spina bifida) is significantly reduced by the consumption of multivitamin preparations around the period of conception (*J.A.M.A.* (1988) **260** 3141–3145). A major double-blind study in England is investigating this issue at the moment. Other studies have shown that zinc deficiencies in the mother can lead to defective developments in the baby's central nervous system.

In the last few weeks of pregnancy it is not advisable for the mother to take large doses of vitamin C, because the baby runs the risk of vitamin C deficiency shortly after birth.

It can be difficult to cover selenium requirements at the best of times, but after a pregnancy the mother's selenium levels may reach rock-bottom and it can take months before this imbalance returns to normal. Two of the reasons for the low selenium levels at the end of pregnancy and shortly after giving birth are, firstly, that the foetus requires extra-large doses of selenium in the closing stages of pregnancy and, secondly, a large proportion of the mother's selenium reserves are to be found in her milk after giving birth. Selenium supplementation may therefore be advisable during pregnancy.

According to a study carried out at Helsinki University Hospital, the average serum selenium levels in women who had recently given birth were only 60 μg l^{-1}, as opposed to normal levels of 80 μg l^{-1}. During the nursing period, selenium deficiency developed further, but a daily supplement of 100 μg of organic selenium was sufficient to ensure that deficiency conditions did not arise.

Zinc requirements also increase during pregnancy because the baby needs zinc for the development of the musculature and the skeleton. In animal experiments it was discovered that zinc deficiency led to defective developments in the central nervous system.

The Swedish researcher, Dr **Sten Jameson,** and a midwife, **Margareta Bjurström,** recently conducted a study over a one-year period, which revealed that one woman in five develops zinc deficiency in the sixth to tenth week of pregnancy, unless a daily zinc supplementation of 20 mg is taken. Zinc supplementation improved the condition of mother and child in this one-year experiment which attracted a great deal of attention. It is on this background that screening for zinc status and possible zinc supplementation is recommended to pregnant mothers, because it is not possible to compensate for the deficiency through diet alone.

It now also seems possible that zinc deficiency during pregnancy and in early childhood might be connected with the later development of dyslexia. Dyslexic children often suffer from zinc deficiency and it would appear that this is associated with abnormal development of the brain.

5.   VITAMINS AND MINERALS FOR YOUNG CHILDREN

Children are nature's most precious gift.

Herbert Hoover (1874—1964)

British children often have very unhealthy eating habits: they eat too much fat (hamburgers, chips) and sugar (sweets, soft drinks, ice cream) and their intake of vitamins and minerals is too low (which sweetened breakfast cereals cannot compensate for). Children's vitamin and mineral requirements are greater, relative to adults, because of their growth.

Many chronic diseases have their origins in childhood, remaining hidden for decades and then breaking out in the fully grown adult. Arteriosclerosis is an example of a disease which has its roots in childhood.

Pediatricians have begun to appreciate the gravity of the situation but we can also do something to avoid later illnesses for our children by ensuring that they receive sufficient of the appropriate protective micronutrients. Healthy food is a good investment for the future of our children.

The newborn baby

Allergies became more common when women began to breast-feed their babies for a shorter period of time. Fortunately breast-feeding is once again more popular. Breast milk contains about 8% polyunsaturated fatty acids compared to only 1% in cow's milk. Breast milk is the only foodstuff which contains significant quantities of the polyunsaturated fatty acid γ-linolenic acid (see 6.2). Some breast milk substitutes contain added quantities of vegetable fats. Some research has suggested that the consumption of these fatty acids in infancy can lessen the later risk of arteriosclerosis. γ-linolenic acid protects the baby from a whole series of disorders: convulsions, infections, allergies and dehydration in warm weather. It also protects against metabolic disorders, obesity and psychological problems. This last-mentioned factor has been established in studies which compared breast-fed babies with babies which had been fed with substitutes for breast milk.

Although breast milk substitutes are manufactured so as to resemble breast milk as closely as possible, the real thing is still the best for the baby. Today only iron is generally added to breast milk substitutes, whereas selenium, copper and zinc supplements should also be included. In Scandinavia zinc has been added to breast milk substitutes since 1987. The baby needs milk (breast milk or milk from a bottle) until it is 9–12 months old, after which time it can begin to eat solid food. Care should be taken at this time to avoid feeding with foodstuffs which contain too much salt or sugar, while pre-sweetened food should also be avoided.

Two percent of all children become allergic to cow's milk before they are three years old, after which the allergy usually disappears. The symptoms include rashes, asthma, rhinitis, infections in the respiratory passages, recurring ear infections, stomach troubles, diarrhoea and vomiting. A few studies have even connected hyperactivity in children with cow's milk allergy. When the milk is excluded from the child's diet, the symptoms disappear, provided other allergens are not present in the diet, such as chocolate, eggs, fish and fruits.

Milk can be replaced by soya-milk, in which case this must be accompanied by supplements of calcium and magnesium. It has been discovered that if an infant's diet contains a high level of soya protein (2 g per kg bodyweight per day) this has a negative influence on the workings of the immune system, increasing the risk of recurring infections. It may be that this is due to the low level of bio-availability of selenium in soya beans and that it is this which gives rise to selenium deficiencies.

In order to build up the bones, a baby needs 58 mg of calcium, 23 mg of phosphorus and 7 mg of magnesium per kg bodyweight per day. Iron requirement is around 10 mg: substantially greater or lesser quantities than this increase the risk of infections. The requirement for zinc is 3 mg and for iodine 40 μg. Selenium requirements for newborn babies and very young children have not yet been established but recent studies suggest that one-year-olds should have a daily selenium intake of about 10–40 μg.

These essential micronutrients are found in breast milk, although, especially in the case of zinc and selenium, deficiencies can arise if no supplements are taken by the mother.

Children of pre-school age

As the child grows its energy needs increase. Calorie requirements can be worked out using this simple formula:

Multiply the age in years by 100 and add 1000.

This gives us the number in kilocalories. After five months the baby's milk needs are about a pint per day (350 kcal or 1459 kJ).

A varied diet is the healthiest for the child. An unbalanced diet, regardless of how healthy the individual components are, can lead to disturbances in growth and development. The implementation of short-term diets, perhaps intended to resolve an allergy problem, are hardly likely to lead to serious deficiencies but they should always be followed up by adequate and safe vitamin and mineral supplementation.

High-fibre diets are also healthy for children but if the fibre content is too high this can mean that the food has a low energy content. On the other hand it is advisable to ensure that overweight and diabetic children consume large quantities of fibre.

It is of the utmost importance that children get sufficient quantities of vitamin A and D. Note that vitamin C increases the absorption of iron which is essential in the prevention of anaemia.

6. VITAMINS AND MINERALS FOR SCHOOLCHILDREN AND ADOLESCENTS

Of all those who have radical opinions, adolescents are usually considered to be the most typical. In fact in my experience they are the most conservative group one can meet.

Woodrow Wilson (1856—1924)

Children and adolescents grow quickly and need a lot of energy; they also need energy for playing and for sports activities. Most children, especially the boys, have good appetites and if their diet is sufficiently varied this supplies them with adequate quantities of the necessary micronutrients, i.e. vitamins, minerals and essential fatty acids. Unfortunately many of them are too keen on junk food: hamburgers, chips and snacks. These products certainly contain a great deal of energy but this is accompanied by an excess of saturated fats and salt. Even amongst top athletes astonishing deficiencies in the mineral balance have been detected.

100 g of chips contains 20 g of fat while 100 g of potato crisps contain 40 g of fat. A single gram of fat provides 9 kcal (38 kJ) of energy. A three-course meal of this sort contains 810 fat-calories, plus the other calories contained in the food. A soft drink contains 25 g of sugar and about 100 similarly 'empty' calories. 'Junk food' is,

however, deficient in the protective and beneficial nutrients. Many children and adolescents avoid the fat in milk but nevertheless love the fatty foods from the chip shop or the Wimpy Bar. Milk actually contains very little fat. In a glass of skimmed milk there are only 2 g of fat.

Parents bear the responsibility for the kind of food which their children eat. Children and adolescents should eat more whole-wheat products, vegetables and high protein foods such as meat, fish, milk and cheese. These foodstuffs are rich in calcium and phosphorus, which are essential in the growth of the bones. Girls need to have an adequate iron, calcium and magnesium intake, especially after puberty when menstruation begins.

The body is unable to use much of the iron which is contained in fruit and whole-wheat products. Iron absorption can be increased by vitamin C which is found, for instance, in citrus fruits.

Many young girls have dietary problems. Approximately 1% of girls under 16 develop anorexia nervosa. This is a very serious condition which has to be treated with medical and psychiatric help. According to a recent study from Sweden, zinc supplementation may restore the appetite and menstruation in anorexia nervosa patients. Many young girls eat too little, even when there is no question of a clinical condition. Nevertheless, even in these cases the vitamin and mineral intake falls to disturbingly low levels, in particular with respect to the calcium and magnesium intakes, which are associated with the growth of the bone mass.

Students, and other young people who have left home, are also exposed to serious risks of vitamin and mineral deficiency. They buy and prepare the food themselves and, because of economic reasons, the quality of their meals often leaves a great deal to be desired.

Although classical vitamin deficiency symptoms are not detected in the young, very low micronutrient levels can, however, lead to tiredness, susceptibility to allergies and infections, despondency, lack of initiative, failing memory and irritation.

A well-planned diet is even more important for children who have to follow a strict diet because of diabetes, gluten allergy or other allergies. In the wintertime, children's and adolescents' diets can be supplemented with fish oils (EPA and DHA, see 6.3) which contain vitamins A and D and are vital for the absorption of calcium.

Bad eating habits have a tendency to be 'hereditary' and they are difficult to break down because people are generally resistant to any kind of change. Nevertheless, it is within our power to adjust these habits in the right direction provided that we initiate a massive information campaign, directed at the parents who can then influence their children — or, alternatively, aimed at the children who can influence their parents.

In early 1988 two remarkable studies were published in Britain on the relationship between vitamin and mineral deficiency and dyslexia and intelligence. **Benton and Roberts** described in *The Lancet* (140–143, 1988) how a daily multivitamin and mineral tablet improved the non-verbal intelligence of a group of Welsh schoolchildren, compared to a control group who were given either a placebo or nothing whatsoever. The experiment was later reported in a TV programme on the BBC. The press has described these results and the experts have debated the results eagerly in an attempt to discredit the study. Another British research group, led by Professor **Naismith**, 'repeated' the experiment with non-positive results, but, in my opinion, a significant flaw in their study arose from the fact that they attempted to draw conclusions after only one month's supplementation, instead of the eight months of the original study.

The other remarkable study came from **Grant** and co-workers in the *British Medical Journal* (**296** (1988) 607–609) and dealt with the relationship between dyslexia and zinc. The average zinc status in 30 dyslexic children was only 66% of the values of healthy classmates, who served as a control group. The study indicates a possible connection between zinc and cerebral development.

My research team has conducted a supplementation study on a group of dyslexic children in Helsinki, most of whom also exhibited very low zinc blood values and a disturbed zinc/copper balance. These children received one daily tablet of an antioxidant cocktail, containing zinc, selenium and vitamins, over a period of one year and their blood tests as well as their psychological development were monitored. After eight month's supplementation we observed a significant improvement, in terms of reading and writing errors, in eleven out of eighteen dyslexic children.

7.   VITAMINS AND MINERALS FOR THE ELDERLY

Nobody is so old that he doesn't believe that he will live one more year.

Cicero (106–43 BC)

The normal recommended dietary allowances for vitamins and minerals are often invalid for the elderly, about whose needs little information is available. Elderly people are often suffering from chronic diseases which may require regular medicine. Diseases and medication often increase lipid peroxidation in the body and this increases the need for antioxidant vitamins, minerals and certain essential fatty acids.

There is an increasing interest in the nutritional status of the elderly, the percentage of whom rapidly increases in Western populations. The topic is scientifically relevant because the recommended dietary intakes of the elderly are less well-defined. As changes in caloric intake, physiological functions, physical activity, and body composition gradually take place with ageing, the needs for nutrients, e.g. vitamins, may possibly change accordingly. The question has hitherto scarcely been put, as to whether or not changes in vitamin requirements will result in changes in vitamin status. There are studies which indicate inadequate intakes and/or marginal to low status of several vitamins for the elderly, while an increasing use of vitamin supplements is observed with age. According to those studies which have been conducted, the average daily food intake for women over 70 years old is 1500–2000 kcal and for men 1900–2400 kcal. Nevertheless, a considerable proportion of the elderly consume less than 1000 kcal per day, which implies that they do not get nearly enough of the protective nutrients from their diet.

Traditionally, it has been asserted that healthy old people do not suffer from any form of deficiency of calories, proteins, minerals, vitamins or trace elements and that food supplements are therefore superfluous. To this I would like to point out that these conclusions are no longer tenable. Most researchers agree that it is difficult to compile a well-balanced diet when the daily energy intake is 1600 kcal or less. Several studies have revealed that vitamin, mineral and fatty and amino acid deficiencies are actually all too common amongst the elderly.

Studies based on questionnaires are unreliable, while dietary studies on the elderly are questionable because nutrients are absorbed very differently from individual to individual and the bio-availability is much lower than in the young. In this connection the overall food consumption only plays a limited role. It therefore becomes necessary to determine micronutrient status on an individual basis with the aid of blood or urine analysis. For example, folic acid deficiency is much more common in the elderly than in young people. Requirements of folic acid have to be determined individually because unsupervised supplementation can conceal symptoms of pernicious anaemia, which is caused by vitamin B12 deficiency.

Elderly people can suffer from poor general health, even when medical examinations and routine blood tests fail to reveal any specific illness. It can be difficult for the doctor to prescribe a specific curative treatment for a condition which is due to undernourishment. In this situation micronutrient tests on the blood might yield the correct diagnosis and treatment.

Two studies, conducted by my research team, on the residents of nursing homes for the elderly in the Finnish towns of Hartola and Lahti, showed that supplementation with certain vitamins, minerals and fatty acids can have a beneficial effect, even though the old people in question do not suffer from a specific clinical deficiency condition (see 2.12). Professor Jørgen Clausen, at Roskilde University, Denmark, has been able to confirm our results by supplementing Danish elderly with antioxidants.

8.   MEDICINE AND FOOD

The doctor administers a medicine, which he knows very little about, to a person whom he knows even less about, in order to cure a disease which he knows nothing whatsoever about.

Voltaire (also a doctor) (1694–1778)

Medicine consumption is high in the industrialized world, but neither the patients nor the doctors are clear about what effects medicines have on the bio-availability of vitamins, minerals and essential fatty acids — or conversely, how the effectiveness of medicaments can be affected by micronutrients. The average European receives five prescriptions from the doctor every year. But does he know which medicines can be taken together with which natural preparations and how they influence each other? Everybody who takes medicine ought to read this section and, if possible, discuss these questions with their doctor.

Does your doctor know that the medicine which he has prescribed for you may increase the formation of free radicals and lipid peroxides? Do you counteract this hazardous effect by taking antioxidants? Although a considerable proportion of the contents of the medicine cabinet at home is unnecessary or even dangerous, this does not mean that we should throw half of the medicine supply out. When medicine is used in the right way, it is an important component in the treatment of illness. Nevertheless, every patient can, and ought to, ask the doctor whether or not this or that specific medicinal treatment is really necessary. More often than not the doctor writes a prescription, not because he or she feels that it is absolutely necessary, but because the patient appears to be waiting expectantly for one.

A regular prescription for the treatment of high blood pressure may be a matter of life and death, but prescriptions of painkillers for common headaches, muscle pains or colds certainly are not. Patients with such minor problems are perfectly at liberty to buy effective over-the-counter (OTC) medicines at the chemist. If the doctor starts to write a prescription for this sort of thing, the patient should feel himself entitled to say no to the doctor, especially if the medicine in question is likely to lead to side-effects.

No medicine is effective if the patient does not even take it. When the doctor writes a prescription, he assumes that the patient automatically goes to the chemist, collects the medicine and then follows the directions on dosage. It would appear, however, that almost half of all prescribed medicines are never collected from the chemist, and, of those which are collected, the patient often takes the medicine when and how it suits himself. Doctors call this poor compliance, and they usually blame the patient for it.

As a matter of fact, the doctor ought to feel that it is his or her responsibility to explain in detail to the patient why he has, in fact, prescribed a particular medicine and what the possible side-effects might be. This would also give the doctor the possibility, and the time, to consider whether or not a prescription is really necessary.

Most of us believe that when we swallow a tablet it passes quickly from the oesophagus to the stomach and on to the intestines. This is not always the case. A capsule, which is taken without or with very little liquid, may get stuck in the oesophagus. In 10% of cases where capsules are taken with liquid the capsules still get stuck in the oesophagus. A capsule which has got stuck rarely comes loose again even if a lot of water is drunk, with the result that it dissolves in the oesophagus. This can result in serious ulcerations with certain medicines, strong pain-killers or antibiotics, for example.

The correct way to take a pill or capsule is to take a mouthful of water first, swallow the tablet, then drink a big glass of water (but not juice or other drinks). Water also has the advantage of speeding up the absorption process so that the medicine arrives more quickly and easily in the blood stream.

Medicines can affect the absorption and the effects of vitamins, minerals and other nutrients in unexpected ways. Conversely, the effect of medicine is often dependent upon what it is taken with — whether this is with food, on an empty stomach or with drinks which can inhibit its effectiveness. For example, the antibiotic, erythromycin, is not well-absorbed when it is taken together with juice, while another antibiotic, tetracycline, is rendered less effective when taken with milk.

The doctor who has prescribed a medicine is often surprised at its lack of effect on an infection and perhaps tries another prescription. This might be avoided if both doctor and patient are aware that all medicine should be ingested together with water. In some cases the non-effect may be due to the nocebo effect (see p. 43).

Some medicines increase the elimination of vitamins and minerals, and, at its worst, this can lead to deficiency conditions. For example, some diuretics may cause potassium and magnesium deficiencies. Other medicines affect the appetite, which can also lead to a deterioration in the patient's nutritional status. This may, in turn, decrease the patient's resistance or intensify unwanted side-effects of the medicine.

Generally speaking, it is best to take medicines together with meals, although a range of medicines are subject to delayed absorption on a full stomach. When medicaments are ingested on an empty stomach they are transported more rapidly from the stomach to the small intestine, where they are absorbed into the blood stream through the intestinal walls and carried to the appropriate tissue where they are to have their effect.

When medicine is taken with a meal it can take 2–5 hours before it reaches the small intestine. Only alcohol and copper are absorbed in the stomach, all other nutrients and medicines are absorbed in the small intestine. The small intestine is several yards in length and food takes many hours to pass through it. High intakes of dietary fibre can absorb a proportion of medicine, with the result that the medicine is sometimes transported out of the organism instead of being absorbed in the bloodstream. Wheat bran, for example, can delay the absorption of the heart medicine, digoxin. The same applies to some minerals, potassium, for example. Certain antibiotics, such as tetracycline, are not adequately absorbed together with calcium, magnesium, zinc or iron because of the formation of certain special organic metal compounds. Milk contains a lot of calcium and this is why tetracycline should never be taken with milk.

Patients who use digoxin can experience arrhythmia if they have a high consumption of calcium. In order to avoid these unwanted effects these substances should be taken at intervals of 2–3 hours.

Medicines and food supplements, which can give rise to stomach irritation should preferably be taken after meals.

The important antioxidant, β-carotene, which inhibits the development of cancer, is poorly absorbed in the bloodstream when it is taken on an empty stomach. Its bio-availability is greatly improved when it is ingested with fats, because β-carotene is fat-soluble. I therefore recommend my patients to take β-carotene together with food or with preparations of fatty acids, enabling a beneficial clinical effect to be achieved at small doses.

A wide range of medicines lead to increased oxygen free radical formation in the organism. These medicines also increase the level of lipid peroxidation which can be measured in the blood or in the spinal fluid. The following is a list of medicines which stimulate the formation of free oxygen radicals:

- antibiotics (e.g. tetracycline)
- anti-epileptic medicines
- nitrofurantoin (for chronic inflammation of the bladder)
- paracetamol
- psychopharmaceuticals (e.g. phenothiazides)
- theophyllin (asthma medicine)
- cytostatics.

At the present time the World Health Organization is planning a multi-centre study in several different countries to investigate the effects of schizophrenia medicines on lipid peroxidation. In my opinion, it is advisable, on taking these medicines, to ensure that an antioxidant supplement is taken to avoid the harmful effects of oxygen radicals.

Every year people are hospitalized as a result of medicine poisoning. Most of these cases are caused by psychopharmaceuticals, heart medicines, sleeping pills, antibiotics, painkillers, cytostatic drugs, anti-epileptic medicines and the combination of alcohol and medicine.

9.   HEALTHY FIBRE

Bread crowns a good meal.

Finnish proverb

Opinions about the importance of fibre for our health have undergone a radical transformation over the last two decades. In the 1950s fibre was still considered to be a totally useless carbohydrate. In 1969 Dr Denis Burkitt was the first to come forth with the claim that cancer of the large intestine and a variety of other diseases were associated with fibre-deficient diets. Dr Burkitt showed that these 'diseases of civilization' were very rare in Africa, where the population has a high fibre intake, and since then researchers from all over the world have been interested in fibre.

There are very few doctors nowadays who do not recommend fibre for diseases in the digestive system: constipation, gall bladder diseases, cancer of the large intestine and the rectum, diverticles in the large intestine, irritated large intestine and haemorrhoids. Fibre is also recommended for the treatment of diabetes, obesity, varicose veins, midriff hernias and heart diseases. Which foods contain the healthy fibres? Can too much fibre be harmful? And what is fibre actually?

Fibres are those parts of plants which are not digested in the gastrointestinal system. They are a kind of waste product and do not in fact provide any nutrition, unlike the absorbable carbohydrates, fats, proteins, water, vitamins and minerals. Fibres are nevertheless healthy in another way because they facilitate the digestive processes and protect the intestines from a number of diseases, which we will describe in more detail in this chapter.

There are two types of fibre: gel-forming fibres and non-gel-forming fibres. Cellulose, hemicellulose, lignin, pectin, and gum are all different varieties of fibre.

Cellulose is a complex molecule (polymer) which is composed of thousands of glucose units (sugar). Cellulose molecules join together to form fine threads of fibre which are 4000–6000 nanometres long and only 4 nm thick. (One nanometre = one millionth of a metre.) In one millimetre, 200 molecules of cellulose can stand end to end, or 250 000 cellulose molecules can lie side by side.

The *hemicellulose* group consists of polymers of other types of sugar molecules (pentoses and hexoses). Xylose, mannose, arabinose, galactose and their uronic acid derivatives are examples of sugars in this group. Hemicellulose, which is found in fruit and vegetables, consists largely of xyloglucans.

*Lignin*s are polymers of aromatic alcohols. They cover and strengthen the cellulose and hemicellulose structures in tree-like growths. Vanillin can also be produced from lignin.

Pectin is a high-molecular-weight sugar derivative of berries, fruits and other plants. In its basic structure it consists of D-galacturonic acid which joins together in long chains to form pectin.

Pectin is used in the manufacture of jelly and jam and in the bakery, medicine and cosmetic industries. Pectin has a very high capacity for absorbing water.

Gum is a tough, water-soluble polysaccharide consisting of 10–30 000 basic units, mainly glucose, galactose, mannose, arabinose, rhamnose and their uronic acid derivatives.

The food industry uses gum-arabic, tragacanth, karaya-gum, kaobi-gum and guar-gum which are manufactured from tropical and subtropical trees and bushes. Guar-gum has become popular as a cholesterol-reducing remedy in various health food products.

Plant gums consists of polysaccharides in seeds and algae. These are used in the food industry as stabilizers and thickeners. Plant gum from some plants is also used as a laxative.

The physical properties of fibre
The influence of fibres on the gastrointestinal system is due to their molecular structure and size, to the relationship between soluble and insoluble components, and last but not least, to the methods and preparation of foodstuffs, in particular the degree to which they are heated.

Young plant components can absorb a great deal of water, but, as the plant gets older, the proportion of lignin, which is water-repellant, increases and the fibres become drier and tougher. Food processing, boiling and frying, and the digestive processes affect the molecular structure of plants, but plant fibres retain their capacity to absorb water.

Fibres from green plants can absorb a lot of water while bran fibres have a low water-absorbing capacity. Some plant fibres form a mucous (gel) which has a high water content. Gel-forming fibres delay the absorption of sugar, water-soluble vitamins and medicines from the intestines. This property is exploited in the treatment of diabetes because fibres even out the absorption of sugar.

At the same time, as the fibres swell up when they absorb water, the mass of the faeces increases and this reduces the passage time of faeces through the intestines. This is how fibres prevent constipation.

Some fibres, lignin in particular, can absorb organic compounds such as gall bladder acids, various medicines and toxins.

Fibre in foodstuffs
There are fibres in all vegetables, but the fibres in meat and fish are not the same as the fibres in plants. Whole grains, vegetables, fruit and berries are all high in fibre content.

The fibres in pectin and gum are gel-forming and water-soluble. These fibres have most influence in the beginning of the gastrointestinal tract, i.e. the stomach and the small intestine. They reduce the re-absorption of gall bladder secretions (bile) from the intestines into the bloodstream and affect the fat metabolism.

Grains mostly contain insoluble fibres like cellulose, hemicellulose and lignin. There are 40 g of fibre in 100 g of wheat bran, 10 g of fibre in 100 g of whole-wheat flour and 3 g of fibre in 100 g of white flour. On average, fruit contains 3 g fibre/100 g, vegetables 2.5–3.5 g/100 g while root vegetables and potatoes contain 1.5–2.5 g/ 100 g. Raw carrots, by the way, contain as much as 5–6 g fibre/100 g.

In Scandinavia, daily fibre intake is divided among various foodstuffs as in the following table (figures are in grams).

Grains	9.0
Root vegetables and potatoes	2.0
Vegetables	2.5
Fruit and berries	4.0
Nuts and cocoa	0.3
Total fibre intake	17.8

It would appear that the minimum daily requirement for fibre is 30–40 g. People on a purely vegetarian diet usually have a daily fibre intake of 40 g, while the average for the rest of the population is less than 20 g. There are many whose daily fibre intake is less than 10 g. In the course of this century fibre intake has fallen by two-thirds. A similar development has taken place in the rest of Europe and in the USA, where fibre intake was also 100 g/day at the beginning of this century.

In Africa, especially in those areas where maize is the staple diet, fibre intake is 50–150 g/day. In these countries the civilization diseases of the West, such as heart problems, thromboses, cancer of the stomach or intestines and obesity, are virtually unknown. In those parts of Asia where rice constitutes the staple diet the fibre intake lies between the European low and the African high, while in Japan the figures are as low as for Europe.

There are two races who live from a diet which contains no plant fibre: the Masai of East Africa and the Eskimos. This suggests that plant-fibre is not so essential to human existence as the foodstuffs which are absorbed in the organism.

The effects of fibre in the organism
- Fibre increases the bulk of faeces in the intestines and thereby speeds up the passage time. People on a normal mixed diet produce 80–160 g of excrement per day, vegetarians produce 225 g, while indigenous Ugandans produce up to 470 g per day. A high intake of fibre can double or treble the excreted mass.
- Fibre delays the passage of food in the early stages of digestion, i.e. in the mouth and in the stomach, and quickly gives an impression of being full. Fibre also slows down the passage of food from the stomach to the small intestine and this influences the absorption of nutrients into the bloodstream. Pectin and gum fibres suppress intestinal activity because the content of the end of the small intestine becomes harder. Sugar and other nutrients are thereby absorbed more slowly and evenly.
- Fibre accelerates the passage time of the excreted mass through the large intestine. When fibre intake with a mixed diet is 20 g, passage time through the

whole of the digestive system is between 60 and 90 hours. Passage time decreases in proportion to fibre intake; in some African peoples it is as high as 30 hours.

- Pressure on the large intestine decreases, and peristaltic movements are normalized (wave-like contraction and release movements in the intestines which push the intestinal contents along).
- Some fibres function as 'antidotes'. They have a capacity to absorb some of the fats and toxins, carcinogens for example, and transport them out of the body.
- Fibre equalizes and stabilizes the degree of sugar absorption. This is of particular importance for diabetics.
- Fibre contains nutrients such as vitamins and minerals.
- Fibre can also inhibit the absorption of certain minerals, although this effect is usually of minimal significance.
- Pectin and gum fibres lower blood levels of cholesterol. This does not apply, however to 'dry' fibres like wheat bran.
- Fibre can result in a bloated feeling due to the production of wind and gas. It is therefore advisable to increase fibre intake gradually so that the intestines become accustomed to the new diet. At the same time it is wise to increase the fluid intake correspondingly.

The use of fibre in disease prevention and treatment
Fibre is not absorbed in the bloodstream. It does not provide nutrients as such. It is nonetheless beneficial for the intestines and we generally feel better when we take sufficient fibre. Fibre helps in the prevention and treatment of the following illnesses:

- constipation
- diverticles (bulges on the intestinal walls)
- irritation of the colon (the end of the large intestine)
- gall-stone diseases (a possible effect)
- Crohn's disease
- prevention of cancer of the large intestine
- diabetes
- obesity.

Constipation is the commonest reason for increasing fibre intake. There are very few who doubt the expediency of fibre in this situation. Vegetarians very rarely suffer from constipation.

The activity of the large intestine is positively affected by insoluble fibres such as cellulose, hemicellulose and lignin. A high-fibre diet increases the excreted mass by one-third and thereby alleviates constipation significantly.

Diverticles are round, pea-sized bulges on the intestinal walls. They are usually found in the large intestine but can also occur in the small intestine. They do not in themselves give rise to symptoms, but have a tendency to become inflamed. This leads to pains in the lower left part of the abdominal cavity, which fade away after bowel movements. The excreta is usually hard and in small pieces. The inflammation is caused by food remnants and intestinal bacteria which are retained in the bulges and the disease is called diverticulitis. There would appear to be a correlation between diverticles and cancer of the large intestine.

Long-term constipation tends to result in the formation of diverticles, which are so common that they are to be found in every third 60-year-old and in every tenth 40-year-old.

In the 1970s this complaint began to be treated with fibre but the results have not been entirely unequivocal.

Nevertheless, it seems that a daily supplement of 15–30 g of coarse wheat bran alleviates the problem by reducing the pressure on the large intestine. Beneficial effects usually appear after a few weeks of regular consumption of wheat bran.

Irritation of the large intestine is also a common complaint on which fibre can have a positive influence, although it is advisable to start cautiously with the new diet. This is because fibre can lead to a feeling of being bloated which can intensify the symptoms. This was the reason why doctors actually recommended a low-fibre diet until the 1960s in the event of an irritated colon. Over time, however, the intestines become accustomed to the extra fibre so that intake can be further increased and the benefit can be reaped. It should be noted that it is advisable to avoid ryebread as much as is possible if one suffers from irritation of the large intestine.

Gall-stone diseases are not affected by fibre to the extent that was previously assumed. The cholesterol levels in the blood are certainly decreased by a high-fibre diet but this does not affect the secretion of gall fluids. Studies on this area are still far from comprehensive, however.

Cancer of the large intestine would appear to be partially prevented by fibre. This has been observed in experiments with animals and in areas of the world where the population lives on high-fibre diets, although there is still disagreement about how this happens. It was earlier believed that this was a result of the capacity of fibre to absorb carcinogens and toxins in bile. This is no longer the accepted view, and it could be that the cancer-preventive effect is due to the shorter passage time in the intestines, which gives carcinogens less time to exert their harmful effects in the intestines. This is at present an area of active research.

Crohn's disease is a chronic gastrointestinal inflammation of unknown origins. The research results suggest that sufferers of this disease should increase fibre intake and reduce sugar consumption in order to reduce the risk of surgical removal of the intestines.

Diabetes is one of the diseases on which fibre has the most obvious beneficial effects. Diabetics on a high-fibre diet have improved blood-sugar levels and decreased insulin requirements. When combined with chromium supplementation the effects are often further improved.

The best fibre for diabetics is the gel-forming guar fibre. This fibre delays and reduces the absorption of sugar by the intestinal mucous membranes. In one experiment the elimination of sugar in the urine fell by half. This effect is caused by the capacity of guar fibre to make the excreta harder, causing passage time in the small intestine to increase. Other fibres, such as pectin, share this property.

The clearest instances of the beneficial effects of guar fibre can be seen in subjects with the adult type of diabetes which does not demand insulin treatment. This type of diabetes often begins in old age and is almost always associated with obesity.

Obesity is one of the most widespread complaints in the industrialized word. Fibre has been used to tackle obesity with varying degrees of success. In the 1950s methyl cellulose was popular, but when controlled studies failed to reveal any clear effect, fibre faded into the background.

Fibre again became popular in the 1970s, however, and today it is difficult to find a slimming cure which does not recommend a high fibre intake. The value of fibre in slimming is due to its capacity to produce rapidly a feeling of being full.

Heart diseases are also associated with dietary fibre. Gel-forming fibres reduce the reabsorption of bile from the intestines and decrease serum cholesterol levels. Barley and oat, in contrast to wheat, have been shown to lower plasma cholesterol. This effect is attributed to the presence of soluble fibre in the form of beta-glucan or cholesterol synthesis inhibition by tocotrienols present in barley.

Eskimos and the members of some Japanese fishing communities actually live from diets which do not contain any plant fibre, and heart disease does not often occur in these people. It is generally accepted that the high content of fish oils in their diets protects them from heart trouble (see also Chapter 6).

An excessive fibre intake can also lead to problems. If one suddenly changes from a low-fibre to a high-fibre diet it can lead to diarrhoea and feelings of being bloated. This kind of discomfort generally passes reasonably quickly and can be avoided by a gradual introduction of the new high-fibre diet.

A high fibre intake, which is not accompanied by a corresponding increase in fluid intake, can cause violent stomach pains or in the worst cases even lead to volvulus. It can therefore be dangerous to use too much dry fibre for, for example, Crohn's disease. Dry fibres, such as wheat bran, should always be soaked, preferably overnight, in water, yoghurt or some other liquid before they are eaten. Generally speaking, however, the problems associated with fibre are less than those associated with a fibre intake which is too low.

In order to maintain a high-fibre diet it is advisable to eat whole grains, rye bread, crispbread and lots of fruit and vegetables. One can of course buy fibre tablets and other fibre products in the health food shops but this is a more expensive approach than increasing the fibre content in our diet.

Fibre increases the absorption of chromium, which, like the fibre itself, has a beneficial effect on diabetes (see 5.4). Fibre can, however, also lead to a certain extent to a reduction, or the inhibition, of the absorption of other minerals. This happens because fibres contain phytic acid which forms chemical compounds with the minerals and thereby inhibits their absorption in the intestinal mucous membranes. This can result in, for example, reduced absorption of calcium, potassium, iron, magnesium and zinc.

This is the reason why phytic acid is extracted from some fibre tablets and other fibre products. Usually, however, a well-balanced high-fibre diet will contain sufficient vitamins and minerals to compensate for this reduction in mineral absorption. A high fibre intake is therefore not usually in itself the cause of vitamin and mineral deficiencies. If, however, one eats a lot of dietary fibre as part of a slimming cure, it is not unwise to combine this with vitamin and mineral supplements.

10.   VEGETARIAN DIETS

Better is a dinner of herbs where love is than a stalled ox and hatred therewith.

Proverbs, chapter 15, verse 17

Vegetarianism is becoming increasingly common in the western world. In other parts

of the world vegetarian food is much more widespread, either on religious grounds or because there is not enough meat to satisfy everybody's needs.

There are many different kinds of vegetarian but common to them all is that all forms of meat, poultry and fish are excluded from their diets. A vegetarian diet usually consists of vegetables, fruit, grains, legumes, nuts, sprouts and mushrooms. A *vegan* excludes all products from the animal kingdom, while a *lacto-vegetarian* excludes eggs but drinks milk and eats other milk products. An *ovo-vegetarian* eats both eggs and milk products, over and above the vegetarian foods. Eggs contain all the necessary amino acids.

In certain rare cases a vegan diet can lead to vitamin B12 deficiency, because this vitamin is only found in significant quantities in meat, fish, eggs and milk products. Porridge made from various grains can be a source of vitamin B12 to a certain extent, but vegans should generally be checked for vitamin B12 levels, and to be on the safe side, should complement their diets with food supplements.

A mixed vegetarian diet usually contains enough iron, but iron from vegetables is less readily absorbed than iron from animal products. Iron absorption can be improved by eating plenty of fruits, vegetables and berries, which are rich in vitamin C.

Soya beans contain a lot of protein, phosphorus, iron, magnesium, calcium and vitamin E, but can only replace eggs and milk to a certain degree. Vegetarians who do not take milk and eggs can therefore run the risk of calcium deficiency. Vitamin D is only found in animal products and deficiencies of this vitamin can lead to serious calcium deficiencies, which can cause rickets in children. Adults who get enough sunlight produce vitamin D themselves.

A vegan diet can lead to deficiencies in the important amino acids, tryptophan, leucine and isoleucine. To avoid this it is wise to eat mixtures of grains, nuts, legumes and seeds, instead of eating them one at a time. This is also a better way to receive a better balance of proteins. A well-balanced vegetarian diet can be just as healthy as the mixed diet which most people eat. A poorly balanced vegetarian diet, on the other hand, can be injurious to the health of children and adults alike.

A new discovery is that vegans in Scandinavia risk selenium deficiencies. Dr **Mohammed Abdulla** (of Karachi, previously of Lund University) has established that the average selenium intake of Swedish vegans is about 10 µg, as opposed to the RDA of 70 µg. The various micronutrient values of Swedish mixed, lacto-vegetarian and vegan diets have been compared in Dr **Alf Spångberg's** book *Kom i Form* (*Get in Shape*). The results are shown in the table on page 89. Note the low levels of zinc, copper, selenium and vitamin B12.

11.   CANCER AND DIET

Behind clouds of worries we see a glimmer of the sunshine of hope.
                                                    Einari Vuorela (1889–1972)

It has been estimated that 35–40% of all cancer cases are caused by insufficient intakes of anticancer micronutrients, i.e. vitamins, minerals and essential fatty acids.

Nutrient	Nutritional content per 1000 kcal			ADI
	Mixed	Lacto-veg	Vegan	
Total lipids (fat) (g)	44.0	37.0	31.0	
Linoleic acid (g)	4.3	8.2	17.0	
Total sterol (mg)	184.0	174.0	211.0	
Cholesterol (mg)	140.0	69.0	16.0	
Protein (g)	30.0	28.0	24.0	
Total carbohydrate (g)	105.0	132.0	150.0	
Saccharose (g)	21.0	8.0	18.0	
Dietary fibre (g)	6.3*	19.0	29.0	30–40
Sodium (mmol)	49.0	43.0	53.0	
Potassium (mmol)	30.0	47.0	56.0	
Calcium (mg)	392.0	429.0	351.0	
Magnesium (mg)	112.0*	191.0	300.0	300
Iron (mg)	6.5	6.6	9.0	10–18
Zinc (mg)	4.7*	5.0*	6.5*	15
Copper (mg)	0.7*	1.0*	2.0	2–3
Selenium (μg)	17.0*	30.0*	5.0*	200
Iodine (μg)	156.0	—	39.0	
Vitamin B12 (μg)	2.7	0.7	0.2	
Total folic acid (μg)	90.0	182.0	301.0	

g=grams, mg=milligrams, μg=micrograms, mmol=millimoles, ADI=Adequate Daily Intake, *=low compared to ADI.
Source: Alf Spångberg, *Get in Shape*, Hålsokostrådet, Stockholm, 1988.

Hundreds of thousands of people have participated in large-scale population studies in several different countries, in which a correlation has been detected between micro-nutrient deficiencies and cancers. At the moment about 30 major intervention studies are in progress in which healthy people and cancer patients are being given supplements of vitamins, minerals and essential fatty acids. The National Cancer Institute in the USA and several other national health services have invested millions of dollars in an attempt to shed light on the possibilities for preventing cancer with micronutrients.

Some researchers are already convinced that cancer patients ought to be given supplements of vitamins, minerals and essential fatty acids. I treat my cancer patients with dietary adjustments and food supplements after having experienced that this leads to improved prognoses.

The incidence of cancer is constantly on the increase all over the world. The single most important cause of cancer is cigarette smoking, to which every third case is attributable. The incidence of lung cancer would fall by 80% if everybody would give up smoking. It has been estimated that alcohol is responsible for 3% of all cancers and that 4% are due to the work environment.

Living habits and the environment influence the risk of contracting cancer in other ways. This has been indicated in various studies which have compared population groups which live in different parts of a specific country and who have different lifestyles. For example, one study showed that the average 35-year-old

Mormon has 44 more years to live. Other 35-year-olds will live on average for another 37 years — i.e. 7 years less! Mormons do not drink coffee or tea, they don't smoke and they are less likely to contract cancer than the rest of the population.

Diet has been linked with cancer for centuries. In the middle ages it was commonly believed that yeast preparations could prevent cancer. As early the 1500s it was recommended that cancer patients should eat cucumber and pumpkin and avoid grilled or fried meat, fish in jelly and raw eggs. Until recently the orthodox medicine world has considered this sort of advice to be pure nonsense.

In the 1980s, research has begun to investigate more closely the possible links between cancer and diet. The newest information has been provided by large-scale population studies and experiments with animals. With the latter it has been possible to establish that the incidence of cancer varies according to dietary changes and that experimentally induced cancers break out less frequently in experimental animals which have been fed with supplements of selenium, vitamin A and vitamin C. In addition, experiments with rats have shown that intestinal cancers occur less frequently when the rats are fed with a high-fibre diet.

Population studies have indicated that cancers of the colon and the large intestine are typical examples of 'civilization diseases', in this case caused by the high intake of fat and refined and low-fibre foods in the western world, although other factors may also play a part in this statistical correlation. It is as yet not entirely clear what this observation implies for human beings as this will require a large-scale population study with a follow-up period of a number of years, which is a long time to wait. I feel that there are already enough indications in the available evidence to justify the recommendation of low fat and increased fibre intakes in our diets.

The same applies to the possible prophylactic effect of selenium in relation to cancer. Animal experiments and population studies comparing low- and high-selenium areas suggest that there is an effect. We ought to consider exploiting this knowledge now, even if it takes twenty years before this phenomenon becomes an established scientific fact.

In recent years researchers have managed to isolate dozens of different chemical compounds which cause changes in the genes and cancers in experimental animals. These so-called mutagenic compounds are also believed to increase the cancer risk for human beings. Unfortunately we are not only dealing with additives which could simply be forbidden. In some cases it may be a question of a substance which occurs naturally in plants or root vegetables as part of their defence against pests or fungi. Several grams of such substances are to be found in a traditional mixed diet.

Furthermore, other harmful chemical compounds arise in connection with the preparation of foodstuffs, in particular in grilled or smoked meats and fish. Even pepper, the nice brown crust on French bread and roasted coffee beans contain carcinogens.

The majority of carcinogens appear when food is heated to over 200°C, although animal experiments have revealed that this can happen at only 100°C.

It would appear that outrageous over-consumption of lipids, in particular, increases the danger of contracting certain forms of cancer, especially cancer of the gall bladder, the prostate, the womb, the colon, the rectum and possibly breast cancer.

Over-eating, and the accompanying high fat intake, accelerate the production of oxygen free radicals which break down and damage the cells. In other situations it can also be more dangerous to increase the intake of polyunsaturated fatty acids than the intake of the lipids in milk and butter. This is because polyunsaturated fatty acids are more easily oxidized and give rise to oxygen free radical production. This is why it is so important to increase the intake of antioxidants, such as selenium, zinc, vitamins A, B6, C and E and β-carotene, when we raise our consumption of polyunsaturated fatty acids. It is also easier to take an antioxidant supplement than to remove the carcinogens from foodstuffs.

Nitrate, which occurs naturally and which is used by the food industry to give a red colour to processed meats, can also cause cancer. It would appear that vitamin C can counteract carcinogenic nitrous amines in the stomach.

The results of studies on the connection between cancer and diet are still so provisional that it is difficult to draw any definitive conclusions. The link between alcohol and cancer of the liver is, however, well-established, and when alcohol consumption is combined with smoking the risk of contracting cancers of the oesophagus, the stomach and the large intestine is multiplied.

Cancers also derive from radioactivity and from certain chemicals present in the work environment. In addition, correlations have been detected between the incidence of cancer and occupation and education. These factors have an indirect relationship on cancer and may be a reflection of different dietary and living habits.

Cancer is almost invariably an acquired disease. That is to say that only very few cases of cancer are the result of inherited genetic characteristics, although the members of certain families may inherit a tendency to contract specific forms of cancer. This might be expressed as an inherited weakness in resistance, caused by the living habits of the family in question.

City-dwellers are more exposed to the risk of cancer than the members of rural communities. This can perhaps be interpreted in terms of the likelihood that the rural population is less exposed to carcinogens present in the urban environment.

There is no way in which we can avoid contact with all carcinogens and it is not necessarily easy to change the way in which we live. Nevertheless, it is possible for us to avoid the most prominent risk factors and to reinforce the immune system against cancer.

How does cancer occur?

There are two steps in the development of cancer: the emergence of the tumour (carcinogenesis) and its growth. Chemical compounds which cause cancer are called carcinogens. Professor **Bruce N. Ames,** from Berkeley University in California, drew the attention of the whole world to the presence of carcinogens in our diet (*Science* **221** 4617 (1983) pp 1256–1264). In his article 'Dietary carcinogens and anti-carcinogens' he placed great emphasis on the role of antioxidants as anti-carcinogens. It is the carcinogens which are responsible for the first step in the development of cancer.

The first phase of a cancer is called the *initiation*. At this point the carcinogenic substance attacks the DNA molecules which contain the genetic information in the cells. DNA can be irreversibly damaged by even a momentary exposure to a carcinogen, although it is possible for cancer to occur without the DNA being affected. A DNA change eventually leading to cancer is called mutation.

Carcinogens do not belong to a specific or uniform group of chemicals. Many different carcinogenic substances exist; Ames mentions 16 different examples of dietary carcinogens in his article.

This form of cell damage occurs continuously in all of us, because we are constantly exposed to carcinogens in our food, in stimulants, from air pollution, from environmental chemicals and from background radiation. The commonest sources of carcinogen-stimulated cell damage are the oxygen free radicals. Fortunately, these in themselves are not usually enough to cause cancer.

The initiation, i.e. the first stage of cancer development, can occur several thousand times in our cells but the cell-repair system repairs the damage time after time until, at some point, the defence fails to function. This chronic initiation, an accumulation of daily, *per se* minimal mutations, resulting from confrontation with carcinogenic substances, may lead to a carcinogenic process. Typical common carcinogens are tobacco smoke, benzpyrene from exhaust gases and nitrosamines. Other initiators include oxygen radicals and radioactive radiation.

The next stage is called *promotion*. In this phase a number of chain reactions take place which result in the cell becoming a cancer cell. The normal defences in the organism, which are responsible for cell division, lose control of this cell and then the cancer cell begins to divide itself uncontrollably. The tumour then begins to form clone mass, in other words an excess of genetically identical cells.

Substances which cause cell proliferation (multiplication) are also promoters. Examples of such factors are arsenic, excessive oestrogen, lack of vitamin A, epoxides and excessive exposure to sunlight. Chronic infections and recurrent traumas may lead to chronic inflammation which is also associated with high proliferation rate, due to continuous renewal of the cells in the tissue in question. Doctors classify the promotion stage as a pre-cancerous situation.

Promotion is followed by *progression*, also known as clonal evolution. In this stage of carcinogenesis, the initiated cell population produces genetically slightly different clones of cells. These cells show a high division rate, as well as a capacity to infiltrate into other tissues and to send metastases (daughter tumours) to other parts of the body. Microscopically, the cells now look individually changed and disorganized, and doctors describe the finding as dysplasia.

During the *progression* the cells are described as neoplastic or pre-malignant as long as they do not infiltrate other tissues. When they do so, they become malignant. Promotion and progression cannot be clearly separated from each other, the process is a continuum, at the end of which a fully developed cancer can be identified.

Cancer prevention
A person who is completely healthy enjoys a natural protection against cancer. The healthy individual can tolerate repeated initiations, without the chain reaction of the promotion phase being set in motion. The body's own defence system, i.e. the immune system and the antioxidant defence system, can neutralize the harmful substances and can even help to repair the damage to the DNA in the cells.

Although our food contains quantities of mutagens and carcinogens, there are a number of factors which protect us from cancer. Fibre in our diet, for example, absorbs fat and carcinogens and transports them out of the body.

Vitamins and minerals can inhibit the emergence, growth and the spreading of tumours, both directly and indirectly (through hormonal influence).

Recent studies suggest that *deficiencies* of β-carotene, vitamins A, C, and E and the trace mineral, selenium, can increase the risks of contracting cancer. Conversely, it would appear that *a sufficient intake of these vitamins and selenium can have an influence in protecting us from cancer.*

Chemoprevention programmes of the US National Cancer Institute (NCI)
The term chemoprevention refers to attempts to lower the risk of cancer by means of nutritional anti-cancer substances, mainly vitamins and minerals. At present there are large-scale population studies in progress, most of which are sponsored by the US NCI. The NCI, founded in 1937, now has a Division of Cancer Prevention with a chemoprevention department headed by Dr **Winfred F. Malone.**

According to Dr Malone, NCI seeks to develop chemoprevention against *initiation, promotion* and *progression.* A list of 600 potential anti-cancer substances has already been worked out. *Anti-initiation* substances are mostly enzymes already present in the body. They are biocatalysts which are able to transform the carcinogenic substances into a water-soluble form and eliminate them from the body before they can cause any harm. Other enzymes help the amino acid glutathione in its detoxifying activity.

Chemical carcinogens are made water-soluble and excretable in two different forms of enzymatic reactions. The first type is hydroxylation, i.e., transport of a hydroxyl (OH^-) group into the potential carcinogenic molecule. The other type of reaction is a combination of the carcinogenic molecule with readily water-soluble molecules.

Unfortunately, hydroxylation often leads to formation of epoxides, which are in themselves highly carcinogenic. Therefore, NCI welcomes substances which would activate the second type of reaction, the detoxifying mechanisms of the body's own protective enzymes, such as glutathione-S-transferase, which quenches the radicals.

Any substance which inhibits pathological cell division may in principle be considered as an anti-promoter. Examples of such substances are *selenium, β-carotene, vitamin A* (natural retinol as well as synthetic retinoids), an anti-oestrogen called tamoxiphen, and a substance called difluor-methyl-ornithin (DFMO).

The chemoprevention of the *progression* is based on the hypothesis that a second genetic damage is required for the promoted cell to turn into a fully developed cancer cell. Principally it would be possible to prevent this second change like the initiation in the first place.

At the time of writing NCI has selected β-carotene and six other potential anti-cancer substances for intervention studies in human populations. These studies are now in progress. Other substances will be tested in the near future.

The test procedure includes five steps:

(1) pre-clinical laboratory research with potential anti-cancer substances in cell cultures;
(2) animal experiments testing the effect of potential anti-cancer substances against certain carcinogenic substances;

(3) general safety and toxicological testing of the substances which have shown anti-cancer properties in the second stage of testing;
(4) clinical studies comprising a limited number of healthy volunteers (phase 1 studies);
(5) clinical intervention studies on large groups of people or on selected risk groups (phase 2 studies).

Pre-clinical tests (i.e. the first three stages of the above listed test procedure) are already being carried out on 75 chemoprevention candidates. In 1988 alone the NCI have spent 5.1 million dollars on these pre-clinical tests.

At this moment, 19 clinical studies are in progress, involving testing of β-carotene, vitamin A, synthetic retinoids, vitamin C, vitamin E, folic acid, calcium, fibre-containing diets, wheat corn and an anti-inflammation drug called piroxycam.

Population studies on cancer and antioxidants
This point is so important that I want to take the opportunity of substantiating it with some studies which have been carried out in this field. In 1972 a nutritional intervention project, sponsored by the World Health Organization (WHO), was initiated in the province of North Karelia in East Finland. The project was originally designed to investigate the reasons for the notoriously high mortality rates for cardiovascular diseases in this area, and to find measures which could lower this high incidence. More than 12 000 healthy subjects were examined and their blood samples were stored for later use in the investigations into the role of selenium in cancers and cardiovascular diseases.

The first study on selenium and cancer comprised of a group of 8113 subjects who were then monitored over a period of six years, from 1972 to 1978. Serum selenium concentrations were analysed in 1983. 128 persons (women and men) between the ages of 31 and 59 years had cancer. One member of the control group was matched to each member of the case group according to age, gender, daily cigarette smoking and serum selenium levels. The average serum selenium concentrations were lower among cases than controls (50.5 as opposed to 54.3 micrograms per litre). For those with the lowest serum selenium values, the relative risk of cancer was increased by more than three times. No associations between cancer and vitamins were studied in this investigation.

The second study from North Karelia involved 12 155 persons who were followed up over four years, from 1977 to 1981. In all, 56 persons died of cancer, and from 51 of them (30 men and 21 women) a stored serum sample was available for selenium analysis. Again, the average serum selenium concentrations among future cancer cases were initially significantly lower than among controls (53.7 as opposed to 60.0 μg l^{-1}). Those who subsequently contracted cancer had 12% lower serum selenium concentrations in the blood samples taken at the start of the study, before the individuals in question actually contracted cancer. Among men the difference was 19% whereas no difference was observed in women cases and controls. Smokers who later developed cancer initially had serum selenium concentrations which were 22% lower than those of their smoking controls.

Serum vitamin A levels were initially significantly lower in cases than in controls (483 versus 524 μg l^{-1}). In male smokers who subsequently died of cancer the initial serum vitamin A levels were 26% lower than among their controls, while the ratio of serum vitamin E and serum cholesterol was 11% lower in cases than controls. Serum selenium levels of less than 47 μg l^{-1} were associated with a 5.6-fold increase for the risk of fatal cancer. Subjects with both low selenium and vitamin E levels in serum were exposed to an 11.4-fold increase in the risk of contracting cancer.

From 1966 to 1972 the Mobile Unit of the Finnish Social Insurance Institution carried out multiphasic screening examinations in various parts of Finland. The original study comprised 62 440 adults. In the mid-1980s a study was designed to investigate the relationship between vitamins, selenium and cancer. A sample of 15 093 women between the ages of 15 and 99 years and initially free of cancer was followed up over a period of eight years. An examination of the National Cancer Register revealed that cancer was diagnosed in 313 of these women during this period. Vitamin E (alpha-tocopherol) was measured from stored serum samples of the cancer patients and 578 matched controls. An inverse correlation was observed between vitamin E and the risk of cancer, even if the cancers in the first two years of the follow-up were excluded. Women with the lowest serum vitamin E values (less than 7.9 mg l^{-1}) had a 1.6-fold risk of contracting cancer, compared to those women with the highest vitamin E values in serum. Low serum alpha-tocopherol values strongly implied an increased risk of epithelial cancers but only a very slightly increased risk of cancers in the reproductive organs exposed to oestrogens.

A significant interaction was observed between serum selenium and serum vitamin E levels with respect to hormone-related cancers, in particular to breast cancer. Subjects with low serum levels of both selenium and vitamin E had a significant, 10-fold higher risk of breast cancer.

The Finnish Social Insurance Institution and the Finnish Cancer Register designed a similar follow-up comprising of 21 172 men, all initially free from cancer. The original blood samples had been taken in 1968–1972 and 453 cancers were diagnosed during the ten years of follow-up. The serum vitamin E levels were measured from the stored samples from these men and from 841 matched controls. The average vitamin E levels among cancer cases and controls were 8.02 and 8.28 mg l^{-1}, respectively. A high serum level of vitamin E was associated with a reduced risk of cancer. The relative risk of contracting cancer for those with the highest serum vitamin E levels was 0.64, compared to those with low vitamin E values. This association was strongest in younger age groups with cancers unrelated to smoking.

Furthermore, the association between serum selenium and the risk of lung cancer was investigated in a longitudinal study of the same group of 21 172 healthy men initially aged between 15 and 99 years. During a follow-up over 11 years, 143 cases of cancer were diagnosed and reported to the Cancer Register. The serum selenium was measured from stored samples from these patients and 264 matched controls, and an inverse correlation was observed between serum selenium concentrations and the incidence of lung cancer. The average serum selenium was $57 \mu\text{g l}^{-1}$ for cancer patients and $61 \mu\text{g l}^{-1}$ for the matched controls. The smoking-adjusted relative risk of cancer was 3.3-fold for those with the lowest selenium values, compared to those with the highest serum concentrations of selenium. This association persisted after adjustment for several confounding factors and were not related to serum levels of vitamin A or vitamin E. They were not secondary to early cancers either, since they persisted after exclusion of cancer cases diagnosed during the first two years of follow-up. This study supports the hypothesis that high intakes of selenium may protect against lung cancer.

The risk of cancer in the upper and lower gastrointestinal tract was investigated in another longitudinal follow-up study based on 36 265 Finnish men and women, all of whom were initially free of cancer. Serum vitamin E and selenium concentrations were measured from blood samples taken at the beginning of the follow-up, in other words, while the subjects were still considered to be healthy. During 6–10 years of follow-up, a total of 150 gastrointestinal cancers were diagnosed and reported to the Finnish Cancer Registry. Blood vitamin E and selenium were then compared with 276 matched controls, i.e. persons from the same follow-up group who did not contract cancer. 'Matched' means that these control subjects were chosen to represent the cancer cases with respect to age, sex, and other usual health parameters, apart from the future diagnosis of cancer.

Subjects with low serum vitamin E or selenium concentrations had an increased subsequent risk of cancer of the upper gastrointestinal tract. The relative risk of upper gastrointestinal tract cancer was 3.3 for men with serum selenium levels in the lowest quintile and 2.2 for men with serum vitamin E levels in the three lowest quintiles, compared with subjects in the higher quintiles. Serum concentrations of vitamin E or selenium in general were not inversely associated with subsequent risk of cancer of the colon (large bowel) or the rectum.

The authors, representing the Finnish Social Insurance Institution, The National Institute of Health and the Finnish Cancer Registry, concluded that high serum vitamin E and selenium levels were associated with a lower risk of gastric cancer (cancer of the stomach) but not of cancers of the colon or rectum. The association was stronger for men than for women. 'The results are in line with the hypothesis that high selenium and vitamin E intake protects against some cancers caused by dietary factors', wrote the authors in *International Journal of Cancer* (1988).

In summary, several epidemiological studies seem to confirm that in a 'healthy' Finnish population there are subgroups with low serum concentrations of selenium, vitamin E and vitamin A and an increased risk of cancer. Two separate studies yielded evidence which suggested that a simultaneous deficiency of selenium and antioxidant vitamins was associated with a 10-fold increase in the risk of cancer, compared to those with higher serum levels of these micronutrients.

These Finnish studies support other follow-up investigations from Sweden, the Netherlands, Switzerland and the USA suggesting that a high nutritional intake of selenium and antioxidant vitamins may be important in cancer prevention.

In England the so-called BUPA study, involving 22 000 men who had delivered blood samples while healthy, revealed that the 271 persons who later contracted cancer had lower blood levels of β-carotene (mean $198 \, \mu g \, l^{-1}$) than those who did not contract the disease (mean $221 \, \mu g \, l^{-1}$).

In Switzerland the Basel study confirmed the same tendency: a low level of β-carotene and of vitamins A, C and E implied an increased risk of contracting cancer. In Sweden 10 000 men between the ages of 46 and 48 years were monitored over a period of 10 years. During this observation period the men with the lowest selenium blood concentrations were more receptive to cancers. Corresponding studies in the United States and Holland have indicated the same tendency: low selenium blood values increase the risk of contracting cancer, and when this is combined with deficiencies of vitamin E and vitamin A, the risk is greatly increased. Many similar studies have been conducted, in which the participants provided blood samples for analysis prior to a later cancer diagnosis. In each case the same trend was observed: low levels of vitamins and/or minerals were a warning of ensuing cancers. It should be borne in mind, however, that epidemiological studies at the very most may reveal existing associations only, not an absolutely certain causal relationship.

Without prospective intervention studies, the question 'Does selenium prevent cancer?' has not yet been fully answered. The inverse associations observed between a nutrient and the risk of cancer may be an artefact due merely to association of nutrient ingestion with some truly protective dietary habit(s) or component(s), or to association of nutrient ingestion with the avoidance of some truly harmful dietary habit(s) or component(s). Consequently, randomized intervention trials with large groups and a long follow-up period are needed.

In Finland a major study is in progress, involving the participation of 30 000 smokers, in an attempt to monitor the possible preventive effects of β-carotene on lung cancer. The costs, about 35 million pounds, are being met by the National Cancer Institute (NCI) in the United States. NCI at present sponsors no fewer than 14 major β-carotene supplementation studies worldwide. In one of them over 20 000 doctors in the USA take 50 mg of β-carotene every other day for several years at a time.

Research in this field is remarkably intensive and we can expect a wealth of new information on the relationship between cancer and diet over the next few years. At the moment there is no definitive diet which can provide 100% protection against cancer but the points listed below represent a summary of our present knowledge on how to reduce the risk of cancer:

- Don't smoke
- Avoid too many calories and obesity
- Don't drink too much alcohol (maximum two drinks per day)
- Avoid too many fatty foods. Fat content in our diet should not exceed 30%

- Avoid salty and strongly spiced food
- Avoid smoked, grilled and over-fried foods
- Get plenty of fibre: eat whole grains, fruit and vegetables
- Take a daily supplement of selenium (100 μg of L-seleno-methionine)
- Eat sufficient β-carotene — either two large carrots per day or 1 to 2 β-carotene tablets (one carrot contains 8000–10 000 IU β-carotene)
- Eat 30 mg of ubiquinone every day
- Take 200–300 mg of vitamin C per day
- Ensure that you get enough vitamin E (200 mg per day)
- Avoid bright sunlight and especially avoid getting sunburned.

Most of this advice is not especially new. The great doctor of ancient Rome, Galen, provided his contemporaries with similar advice for avoiding cancer, although it is unlikely that he knew a great deal about selenium or vitamins.

Note that I do not recommend here any megadoses of vitamins and minerals but rather a modest supplementation in order to ensure safe and adequate daily intake of anti-cancer antioxidants.

People who come from families where cancer cases have been common, and who therefore may have a certain hereditary tendency to contract the disease, ought to pay special attention to the above guidelines. It should be remembered that it is never too late to make a start on cancer prevention, because cancer usually has a very long development period. It is not the number of initiations which cause cancer to break out, but the promotion phase and it is precisely this phase on which antioxidants like β-carotene can have a prohibitive effect.

Cancer prevention ought to begin in childhood. Children should learn healthy eating habits at an early age. Nursing mothers should ensure that they get a sufficient intake of vitamins and a daily supplement of 100 mg of selenium. A modest antioxidant supplement for children over the age of six years is also advisable.

Biological monitoring of antioxidants — a new approach to cancer prevention

Drs **Fred Gey, Georg B. Brubacher** and **Hannes B. Stähelin** from Switzerland recently reviewed the relationship between plasma levels of antioxidant vitamins in relation to cancer (and coronary heart disease). They proposed a completely new approach to detecting antioxidant vitamin deficiencies which may predict the future risk of these common diseases in the population. Based on the present data on epidemiological studies, they published (in the *American Journal of Clinical Nutrition*, 1987) desirable plasma levels and the corresponding daily intake of the most important anti-cancer vitamins, as delineated in the following table, which shows present evidence for plasma levels of essential antioxidants associated with lessening risk of major health hazards and suggest prudent daily intake (PDI) to achieve potentially protective plasma levels in middle-aged non-smoking men.

I would like to propose the desirable selenium concentration in whole blood, which has been shown to correlate with reduction of lipid peroxides in human serum, namely 200–350 μmol l−1. Enhanced lipid peroxidation is now also recognized as a factor increasing the future risk of cancer and cardiovascular diseases. The appropriate daily intake of selenium which results in a concentration of 200–350 μmol l^{-1} is in the order of 200 μg of selenium in the diet (or diet plus supplement pill).

Antioxidant	Health hazard	DPL	PDI
β-carotene	Lung cancer Lung diseases Stomach cancer	>0.4	15 mg (minimum in NCI sponsored intervention trials)
Vitamin A	Stomach cancer	>2.1	1 mg (RDA USA, Canada, etc.)
Vitamin C	Ischaemic heart disease	60 mg	(RDA USA, Canada, etc.)
	Intestinal cancer	>23	
	Other hazards	>50	100–150 mg
Standardized vitamin E	Ischaemic heart disease	>25	30–50 I.U.
	Intestinal cancer	>30	60–100 I.U.

DPL–Desirable plasma level (μmol l−1); PDI=Prudent daily intake for optimal health; NCI=National Cancer Institute of the United States.

It is important to note at this juncture that there are many countries where the natural selenium intake is in the region of 300–400 μg per day. There are no signs that anyone has suffered from this higher level of selenium intake.

LITERATURE

Gey, K. F. On the antioxidant hypothesis with regard to arteriosclerosis. *Bibl. Nutr. Dieta.* **37** (1986) 53–91.

Gey, F., Brubacher, G. B. and Stähelin, H. B. Plasma levels of antioxidant vitamins in relation to ischemic heart disease and cancer. *Am. J. Clin. Nutr.* **45** (1987) 1368–1377.

Hurley, L. S., Keen, C. L., Lönnerdahl, B. and Rucker, R. B. (eds). *Trace elements in man and animals* **6**, Plenum Press, New York – London, 1988, p. 724.

Hayaishi, O., Niki, E., Kondo, M. and Yoshiwaka, T. (eds). *Medical, biochemical and chemical aspects of free radicals.* Vols 1 & 2, Elsevier, Amsterdam — New York — Oxford — Tokyo, 1989, p. 1559.

Tolonen, M. Finnish studies on antioxidants with special reference to cancer, cardiovascular diseases and ageing. *Int. Clin. Nutr. Rev.* **9** 2 (1989) 68–75.

4

Vitamins

1. NEW DEVELOPMENTS WITH VITAMINS

Nothing is so established that it is impossible to say something new about it.
Fjodor Dostojevski (1821–1881)

Vitamins are every bit as relevant today as they were almost 100 years ago, when they were first discovered. We can divide the last century into three periods to show how we came to appreciate the immense significance of vitamins. First came the discovery of vitamins and their connection with the classical deficiency diseases: scurvy and vitamin C, rickets and vitamin D, and many others. Later, 25 new diseases were discovered in human beings in the twenty years between 1954 and 1974. These diseases are all forms of metabolic disturbance which can be treated with vitamins at 10–10 000 times the normal recommended doses.

Since 1980 previously unknown properties of vitamins have been discovered. Certain vitamins are antioxidants, i.e. they protect the cells from rancidification. Vitamins strengthen and make more effective the functioning of the immune system, which is responsible for neutralizing invading microorganisms, such as bacteria, viruses and carcinogens. It now also seems possible that vitamins can have a delaying influence on the ageing process and inhibit senile decay. β-carotene and vitamins A and E, in particular, would appear to have a prophylactic effect on cancer and can perhaps also be used as supplementary treatment for this disease. Vitamins can also have a positive preventive effect on damage to the central nervous system in embryos and children.

Vitamins, minerals and essential fatty and amino acids can counteract, co-operate with, or reinforce each other. It is therefore extremely important that we know how they function. Our knowledge of how all vitamins work is far from complete but in this section I have attempted to provide an overview of our present knowledge about these vital micronutrients.

What are vitamins?

The word 'vitamin' was invented in 1912, when the Polish biochemist, Dr **Casimir Funk**, isolated the first vitamin while working in London. It was a substance which occurred in rice husks and which could prevent the dreaded disease, beriberi. Funk thought that this substance was an amino acid and called it a vitamin. *Vita* means 'life', so vitamin means the *amine of life* and this is still the same term which we use today.

Vitamins can be defined in the following way: *Vitamins are organic compounds which the organism requires for metabolic processes. The body is not in itself capable of producing sufficient quantities of vitamins. The different vitamins are neither chemically nor functionally related to one another. Each vitamin has its own function in the organism and cannot be replaced by any other substance.*

It is possible for an organic compound to be a vitamin for a specific species of animal or for human beings and yet be of no significance for another species. For example, vitamin C is a vitamin for human beings because we are unable to produce it ourselves. Many other animal species do produce vitamin C in the body and it can therefore not be considered as a vitamin for these species.

The human organism can produce ubiquinone, niacin, and vitamin K, although not in sufficient quantities. Under the influence of sunlight we manufacture vitamin D in the skin — this vitamin has its effect in the intestinal system and the bones, i.e. fairly far removed from the place of production — and in this sense vitamin D can be described as a hormone.

The discovery of vitamins

For thousands of years, human beings have known about diseases which could be cured with certain foods. In ancient Greece, Rome and Arabia, goose liver was used in the treatment of night-blindness. **James Lind**, a Scottish doctor in the British Navy, discovered in 1757 that scurvy could be prevented and treated with fresh fruit. The disease was a serious problem for long-distance sailors and it often happened that the whole crew of a ship died of scurvy after they had been at sea for several months at a time. After James Lind's discovery, lemons and orange juice became compulsory for sailors in the Royal Navy, and scurvy ceased to be a problem. Today we know that these two diseases, night-blindness and scurvy, were cured by vitamin A and vitamin C respectively.

Research into the vitamins themselves first began at the beginning of our own century. The English biochemist, **F. G. Hopkins**, conducted experiments which involved feeding experimental animals a diet consisting of pure carbohydrates, proteins, fat, minerals and water. Despite this, the animals were unable to procreate and instead became ill and died. On the other hand, they thrived when they were fed on an unprocessed natural diet.

The experiments were therefore able to show that the refinement of foodstuffs removed some essential dietary factors. It was not known, at that time, what these factors were, but it was appreciated that we only needed tiny quantities of them — in the range of micrograms or milligrams. The significance of these substances is nevertheless far greater than their modest quantities might suggest.

The fat-soluble vitamin, vitamin A, was discovered in 1913–1914. Prior to this, vitamin B had been discovered (the letter B was used to denote beriberi, the disease which the vitamin could cure). At this stage researchers had only reached the first two letters of the vitamin alphabet.

Between 1910 and 1930 a whole new series of vitamins were discovered and they were named in alphabetical order. The anti-scurvy vitamin was called vitamin C, vitamin D could prevent rickets, vitamin E was discovered to be essential to fertility and vitamin K was necessary for the coagulation of the blood. The letters F, G, H, M and P have also been used to denote vitamins or vitamin-resembling compounds but these terms are no longer in use.

As researchers gradually got to know the chemical structure of the various vitamins, it became possible to synthesize them. At the same time the different vitamins were given chemical names which are more and more commonly used today. Vitamin B1 became thiamine, B2 is riboflavin, B6 is pyridoxine, vitamin C is ascorbic acid, vitamin E is tocopherol and so on.

Up until now there are 13–14 known vitamins, of which four are fat-soluble and the rest are water-soluble. The actual number of vitamins depends on whether or not we include organic compounds, such as choline, inositol and para-aminobenzoic acid, which resemble the B vitamins in their structure and their effects. The status of vitamin B15 or pangamic acid and vitamin B17 or amygdalin (laetrile) is especially disputed.

Most vitamins are in reality a group of related compounds. A good example of this is vitamin B6, which is a common term for pyridoxine, pyridoxal and pyridoxamine. The effects of the three compounds in the organism are identical and they therefore represent a functional unit. Another example is vitamin A and its precursors, the carotenes. These compounds also function in an identical fashion in the body with one crucial difference: an overdose of vitamin A can cause poisoning while β-carotene is non-toxic. But we can also classify β-carotene as an independent vitamin, because, unlike vitamin A, it can react with the free radical, singlet oxygen.

How do vitamins work?

Each vitamin has a unique task in the organism. Vitamins are participants in enzyme systems, or so-called coenzymes, and they catalyze, i.e. set in motion, certain chemical reactions in the food metabolism. Vitamins are essential to the carbohydrate, protein and lipid metabolisms. Most vitamins cooperate with each other and also with a number of minerals and fatty acids. They are therefore synergistic, i.e. they reinforce each other.

This cooperative property of vitamins and minerals means that substances which are taken together have a stronger effect than substances taken by themselves. This discovery is relatively new and has been instrumental in the development of new methods of treatment. Essential fatty acids are also participants in this synergistic process.

We now find ourselves in the third phase of the history of vitamins. The first phase incorporated the discovery of vitamins and their corresponding deficiency diseases, the isolation of the various vitamin compounds and their synthetic manufacture. In the next phase a whole series of new diseases, which were caused by vitamin deficiencies, were discovered and more or less eradicated.

The third phase in the fascinating history of vitamins began in the 1950s, when a number of doctors started to treat patients with larger doses than the more conservative RDAs. Since 1980 a range of new and previously unknown aspects of vitamin deficiency have come to light — and in many cases these deficiencies are of such a modest degree that they would not even have been registered 15 years ago.

These deficiencies appear to be widespread, and they seem to be associated with a number of common diseases like cancer and ageing itself. We have also begun to understand the nature of the synergistic relationship between vitamins and minerals, something which is opening up whole new perspectives for a fourth phase in the history of vitamins.

How much do we need?

A great deal of our present concern for our health is associated with the developments which have taken place in the food industry, such as processing, refining and the increasing use of additives. Previously foodstuffs were consumed close to the place of production, on the farm or in the village, so everybody could see how they were cultivated, bred or processed. In general, methods of cultivation and breeding were more natural in the past than they are now, where we use huge quantities of fertilizers and pesticides, and where livestock are treated with all sorts of medicines. Foodstuffs are then processed in factories so we have no way of knowing what actually happens to them before they arrive on our table.

We are perhaps closer to the situation that Hopkins, Funks and their experimental animals were in at the start of the century — when foods are processed and refined the essential vitamins and minerals disappear.

Healthy people on varied diets, who also eat plenty of fruit and vegetables, presumably consume enough of most vitamins. On the other hand, requirements are often much higher than normal, for example, for pregnant and nursing mothers, smokers, and those who suffer from general poor health or chronic illnesses. An unvaried diet or malabsorption can also lead to vitamin deficiencies. Environmental pollution alone increases the need for certain antioxidant vitamins. I have recently met Japanese professors of nutrition who themselves take 400 mg of vitamin E (as opposed to the RDA of 10 mg) as protection against heavy pollution.

The widespread use of synthetic medicines is another important factor in the extra formation of free radicals in the body, which have to be counteracted by antioxidant vitamins.

Smoking, alcohol, contraceptive pills, coffee, antibiotics, other medicines (e.g. psychopharmaceuticals, anti-epilepsy medicines, theophyllin) and industrial chemicals can seriously impair absorption capacity or impose increased demands on antioxidant intake.

Population studies have revealed that vitamin deficiencies are far more widespread than was previously assumed. Many people do not get enough β-carotene, ubiquinone, or vitamins A, B6, C, and E. Their intake is not so low that they are hit by the classical deficiency diseases, but it is nonetheless low enough to make them more susceptible to other symptoms and diseases, which could otherwise have been avoided or postponed until later in life.

Treatment with vitamins

Vitamins are nutrients; they are not foreign molecules to the body. It is therefore much safer for us to use vitamins than other forms of medicine. Vitamins can also be used quite safely in the prevention of disease if we follow the instructions for use. Even when large doses of vitamins are employed in treatment there is fortunately a very large gap between the safe and the toxic dose. Generally speaking, it is safe to take 10 times the recommended daily doses of fat-soluble vitamins and 50 times those of the water-soluble vitamins.

Doctors call this gap the therapeutic margin. Vitamins have such a large therapeutic margin because, unlike medicines, they are substances which the organism is already familiar with. Nevertheless, there is a palpable risk of overdosing and this can have serious consequences. It is known that the fat-soluble vitamins, A, D, E and K, are stored in the organism and that long-term overdosing can be harmful. A careless and excessive overdosing of the water-soluble B6 vitamin can also lead to problems. Mention of vitamin poisoning fortunately occurs only very rarely in the medical literature. In Europe we are used to smaller dosages than in America, and this is probably why I have never seen any references to vitamin poisoning in Finland. Nevertheless, it is advisable to keep a close eye on our intake of these substances.

2. WATER-SOLUBLE VITAMINS

What is wrong with facts is that there are too many of them.
                                                     Samuel Crothers (1857–1927)

Vitamin C and the B group vitamins are water-soluble. In this chapter I will describe briefly the many aspects which they have in common: how they are lost in the preservation and preparation of food, about their absorption and elimination, about deficiency conditions and about the effects of these vitamins in the organism.

Water-soluble vitamins are mostly to be found in bodily fluids, in serum and in the liquids between cells. When serum is saturated with a water-soluble vitamin, the excess is eliminated with the urine. The vitamin content of serum also reflects upon the vitamin content in the enzymes and in proteins. Water-soluble vitamins are not stored to any appreciable extent and it is therefore important that we receive a daily supply of them. A discontinuation of vitamin intake does not lead to a serious deficiency condition until after a period of several months — as was the case with the unfortunate victims of scurvy caused by vitamin C deficiency.

Generally, the water-soluble vitamins are harmless for the organism even in large doses because the excess is eliminated with the urine. Megadoses (i.e. several grams) of vitamin C and vitamin B can lead to stomach problems. The highest dietary concentrations of water-soluble vitamins arise in fresh fruit, vegetables, root vegetables, meat and milk products. This is why a vegan diet, which excludes meat, milk products and eggs, can lead to vitamin B12 deficiency.

Vitamins are not utilized by the organism in the form in which they arise in food. They must first be transformed into what are known as co-enzymes. This transformation process functions very poorly in some people and it is not unusual to find deficiency conditions where the actual vitamin intake in itself is sufficient. This kind of deficiency can, however, usually be remedied or prevented with very high doses of vitamins.

Today we are very well-acquainted with the function of water-soluble vitamins in the organism and only the coenzyme of vitamin C is still unknown. All the water-soluble vitamins have inter-connected functions and this is why a deficiency of one of them can result in the others not being utilized in metabolic processes.

There are a number of medicines which suppress the activity of B vitamins in the cells. The result of this is that the enzyme, in which the B vitamin in question participates, loses its effect. Certain medicines thus act as antivitamins, or opponents of vitamins. Cytotoxins, which are used in cancer chemotherapy, and many antibiotics are examples of such antivitamins.

Since the water-soluble vitamins are almost all found in the same foodstuffs, deficiencies of a single water-soluble vitamin are rarely seen. This usually only occurs when the individual in question has a reduced capacity for absorbing a specific vitamin.

Deficiencies of these vitamins are first observable in rapidly growing tissues with an active food metabolism, such as the skin, hair and nails, the blood, the digestive organs and the nervous system. Symptoms of this sort of vitamin deficiency are anaemia, stomach trouble and nervous disorders.

Water-soluble vitamins are sensitive to light and high temperatures. Vitamins are dissolved in boiling water so it is advisable to cook vegetables and fruit in as little water as possible in order to retain the vitamins — or else use the left-over vegetable water in the food. Vitamin C is, however, already destroyed by being heated up.

Processing of grains removes several important nutrients. This is why it is common practice to add a number of B vitamins to cereals, flour and grains, such as vitamins B1 (thiamine), B2 (riboflavin), niacin and B6 (pyridoxine).

3. FAT-SOLUBLE VITAMINS

The thirst for knowledge is like the thirst for wealth, the more one possesses, the more one craves.

Laurence Sterne (1713—1768)

The fat-soluble vitamins — β-carotene, vitamins A, D, E, and K and the vitamin-resembling ubiquinone — differ widely from water-soluble vitamins in a number of ways. The best sources of these vitamins are vegetable fats and oils, vegetables and the fats in meats, butter and eggs.

Fat-soluble vitamins participate in the metabolism of nutrients together with lipids. They are absorbed together with fat, and if fat absorption is disturbed, this can lead to reduced bio-availability for, or outright deficiencies of, these vitamins. In the blood they bind themselves to chylomicrons and lipoproteins which transport them to the individual tissues. Fat-soluble vitamins are stored in the liver and in fat tissue, from where the organism collects them according to its needs. They are usually eliminated with the excreta and with the bile. Unlike the water-soluble vitamins they are not eliminated with the urine.

It can therefore occur that vitamins A and D (but not β-carotene) can be stored in such large quantities that they can become toxic. A number of cases have been observed where vitamin poisoning occurred in connection with vitamin A and vitamin D, while vitamin E only represents a serious risk in the form of injections. Toxicity has never been observed with tablet consumption of vitamin E in doses of up to 1000 mg. In the USA and in Canada daily supplements of vitamin E of up to 3400 mg have been administered for 10 years to patients with Parkinson's disease, without adverse effects. In theory it is also possible to overdose with intramuscular injections of vitamin K but this is extremely rare.

Vitamin A or D poisoning is usually associated with very high doses over long periods. It is therefore advisable, when considering a treatment with very high doses of fat-soluble vitamins, to seek the advice of a doctor.

Deficiencies of fat-soluble vitamins are often observed in new-born babies who have not yet had the opportunity to store these vitamins in the body. On the other hand, overt deficiency diseases are rare in adults. They are usually observed in connection with a disturbance in the capacity to absorb lipids in the intestines, or a blockage in the bile ducts, which in turn decreases the absorption of the vitamins. Low blood levels of vitamins A and E are most commonly observed in cancer patients.

A connection between vitamin A overdosing during pregnancy and damage to the foetus is often talked about. This area has been thoroughly examined by Dr **Barbara Underwood** in the United States. According to her research, only 18 cases, in which overdosing with vitamin A has been linked with damage to the foetus, have been observed, all of them in the USA. None of the cases are well-documented but in the suspected occasions the daily vitamin A consumption of the pregnant mother was more than 20 000 IU. According to Dr Underwood, vitamin A supplements of up to a daily level of 8000 IU are completely safe, even during pregnancy and breast feeding.

β-carotene has never been known to cause embryo damage in animals or human beings, even though it has used since the 1960s in daily doses of 200 mg over extended periods. Neither are there reports of β-carotene toxicity on humans. A normal supplement of β-carotene contains 6–20 mg per day.

Neither has the vitamin-like substance, ubiquinone (coenzyme Q10), been known to cause poisoning, although three out of every thousand who take a daily supplement of 90 ml or more experience stomach trouble or palpitations. The usual daily supplement is 30 mg which is unlikely to cause any side-effects.

4. VITAMIN DEFICIENCY DISEASES

Who has made the greatest discovery? Chance.

Mark Twain (1835–1910)

Since 1954, 25 different diseases have been discovered which are caused by failings in the body's vitamin functions. These diseases, which attack babies and young children, are admittedly rare and often hereditary. However, they can often be treated with very high doses of vitamins which are 10—1000 times more than the normal requirements.

If the disease is discovered at a sufficiently early stage and treated correctly, the child can look forward to a completely normal life. If the disease is not treated, it leads inevitably to restricted development or even death. The discovery of these hereditary diseases indicates that vitamins may play a previously unknown role in the human food metabolism.

Congenital disturbances in the vitamin metabolism

The cases referred to above are not related to any of the classical deficiency symptoms. These diseases are rare because both parents must be carriers of the relevant genes before the disease breaks out. One case of these symptoms occurs for every 40 000 births. On the other hand, it is not nearly so unusual that the child is born with these genes from only one of the parents and thereby becomes a carrier of the disease itself. This happens to one child in every 200.

The disease appears shortly after birth or in infancy. If these children are not treated with large doses of the appropriate vitamins at an early stage it can lead to irreparable damage, such as mental deficiency, schizophrenia, underdevelopment or organic heart disease.

No new diseases of this type have been discovered since 1974. Although these diseases are rare it is worth bearing them in mind if we detect any inhibition in the mental development of infants.

Reduced vitamin function

The concept of reduced vitamin function is quite new in nutritional medicine. Patients in this category are not as ill as those described above but neither are they completely healthy. So far only one illness is known of this type, Wernicke's syndrome. This is a neurological disease, which is most frequently observed in alcoholics, although it can also occur as a result of undernourishment. It is connected with inadequate thiamin (vitamin B1) activity. If the patient eats normally and avoids alcohol, thiamin functions optimally, but under the influence of alcohol thiamin requirements increase drastically.

5.   VITAMIN A

Every reform originally has a private purpose.

R. W. Emerson (1803–1882)

Vitamin A and its precursor, β-carotene, may prevent cancer. Smokers and the elderly are particularly susceptible to deficiencies of these vitamins. Vitamin A can protect us against night-blindness and defective eyesight and is also essential to the normal development of the foetus and young children. Conversely, too much synthetic vitamin A derivatives can actually cause damage to the foetus. This is why I always recommend β-carotene. Vitamin A is also responsible for the condition of the mucous membranes, it can prevent infectious diseases, it protects the hair and the skin, and β-carotene protects us from the harmful effects of excessive sunlight.

The highest concentrations of vitamin A are found in animal fats and egg white. Carrots and green vegetables are good sources of carotene.

Vitamin A is actually a whole family of compounds, including retinol, retinal, retinoids, carotene and carotenoids. Retinol is the actual vitamin A and presumably has a hormonal function. Retinal is a precursor to the visual pigment, rhodopsin, while retinoids is a common term to describe all the retinols, i.e. inclusive of both naturally occurring and synthetic compounds.

The carotenes are precursors to vitamin A and are found in carrots, among other things. One carrot yields β-carotene corresponding to 8–10000 IU of vitamin A. Approximately 10–40% is lost in cooking, depending on the method of cooking. Carotenes are transformed into retinol, i.e. the actual vitamin A, in the organism. Six molecules of β-carotene remain in the body for every one of vitamin A. 6 μg of β-carotene correspond to 1 μg retinol (vitamin A) which also equals one 'retinol equivalent'. In terms of international units (IU) one retinol equivalent equals 3,33 IU retinol (vitamin A) or 10 IU β-carotene. In other words, 6 mg β-carotene equals 10000 IU β-carotene and 3.333 IU retinol (vitamin A). It is safer to use β-carotene because even very large doses do not lead to the toxicity symptoms or embryo damage which have been observed with excesses of vitamin A.

Vitamin A requirements and metabolism
Actual vitamin A only occurs naturally in animal products. Good sources are cheese, butter, eggs and fish. In the average diet, three-quarters of the requirements for this micronutrient come from actual vitamin A (in the form of retinyl palmitate), while the rest is derived from β-carotene.

Vegetarians cover all their vitamin A requirements with carotenoids. The RDA for adults of vitamin A is 5000 IU, which is about 1000 μg. In Finland the average β-carotene intake from diet is about 2 mg.

Both vitamin A and β-carotene are absorbed in the small intestine together with fats, and this requires the presence of bile and pancreatic secretions. Vitamin absorption is diminished by disturbances of lipid absorption, for example, in gluten allergy, diarrhoea or liver diseases.

Recent studies conducted at the National Cancer Institute in the United States have shown that simultaneous intakes of fats increase the bio-availability of β-carotene.

Very large quantities of vitamin A can be stored in the liver. Zinc is required to release the vitamin from the liver to other parts of the body where it is required. The body can therefore be supplied with sufficient quantities of vitamin A, even when intake is decreased, as long as reserves are ample. The half-life for vitamin A in the human organism is 200–300 days, which implies that the vitamin A balance is not disturbed by minor fluctuations in intake. If vitamin A intake is too low over a period of six months, however, the risk of deficiency can arise. Total stored mass of vitamin A has been estimated at around 1 000 000 IU for adults. Of this, 90% is located in the liver, 1.5% is found in the blood while the remainder is stored in the other tissues. Storage capacity is significantly reduced in liver diseases.

In contrast to vitamin A (retinol), β-carotene blood values return to their original levels within 20–30 days after cessation of supplementation.

The effects of vitamin A and β-carotene
Vitamin A (retinol) *per se* is not a strong antioxidant, like its precursor β-carotene, but retinol has many other vital functions. The importance of vitamin A for good eyesight, fertility, growth, and the optimal functioning of the skin and the mucous membranes has been well-documented for several decades. Vitamin A and its precursors, the carotenoids, have proved to be effective in the treatment and prevention of a number of skin diseases including photosensitivity (sun allergy).

Vitamin A and β-carotene can protect the cell membranes and other cellular structures from the damage caused by oxygen free radicals. The precursor of Vitamin A, β-carotene, is an antioxidant, like selenium, zinc and a range of other vitamins, which can inhibit excess oxidation of fats (lipid peroxidation) in the cells. Exactly how β-carotene functions within the cells is not yet known but it would appear that it happens via the agency of the cells' genetic material, the RNA and DNA. β-carotene has a specific affinity to an oxygen-derived free radical called singlet oxygen. Since retinol (vitamin A) lacks this property, we may regard β-carotene as an antioxidant vitamin in its own right, as distinct from vitamin A (retinol).

The first sign of vitamin A deficiency is deteriorating eyesight in the dark, known as night-blindness. Later developments of this deficiency include further deteriorations in eyesight, dryness of the eyes, dry skin and weakness in the cornea.

β-carotene protects the skin from excessive sunlight. This may result in yellowing of the skin which, in fact, represents no toxic or harmful overdose of β-carotene.

Low blood levels of vitamin A and β-carotene (less than 400 nanomoles per litre) may be a risk factor in cancer, while doses which are slightly larger than the present RDAs would appear to have a prophylactic influence on cancers. The RDAs were set before the anti-cancer properties of these vitamins became known, so the RDAs do not consider the prevention of cancer. This preventive effect is apparently reinforced when vitamin A is combined with sufficient intake of vitamin E and selenium. More recently, it has been discovered that vitamin A may have a supportive influence on the immune system. According to Dr **Alfred Sommer**, in the Third World, vitamin A deficiency lowers infant survival by three to nine times due to impairment of the immune system. This can perhaps explain why vitamin A can have a beneficial effect on long-term and chronic diseases and serves as an illustration of how accepted medical facts undergo rapid revision.

Interest in the anti-cancer effects of β-carotene, vitamins A and E and selenium has grown dramatically in recent years. It is perhaps time to take a closer look at the RDAs for these substances, where re-evaluations for cancer patients and individuals who come from families in which the incidence of cancer is high may be of some benefit.

More than 20 epidemiological studies consistently indicate that a low serum β-carotene concentration predicts future risks of cancer. The risk ratio in these studies varies between 1.5 and 2.5, in other words people with low β-carotene levels run a 1.5–2.5-fold risk of contracting cancer, compared with those who have higher β-carotene levels.

One of the largest studies in this area has been conducted in Great Britain. This was the so-called BUPA study reported by Dr **N. J. Wald** and his colleagues in 1988 (*British Journal of Cancer*). They analyzed serum β-carotene from 22 000 healthy men, aged 35–64 years. Subsequently 271 of them developed cancer. In a comparison between these individuals and the healthy controls from the same database, the serum β-carotene levels in the 'cancer candidates' were initially lower, often 5 years before the cancer was diagnosed. Smokers who subsequently contracted cancer had the lowest levels of β-carotene. Serum β-carotene correlated inversely with the number of cigarettes smoked daily; the more cigarettes smoked, the lower the β-carotene concentrations in the blood and the greater the risk of cancer.

This was the fourth major population study since 1984 showing that blood β-carotene samples, taken before the diagnosis of cancer, predict the future risk of contracting the disease. These four studies involved a total of 769 individuals who contracted cancer after the blood sample was taken.

At the moment several major studies are being conducted on the possible prophylactic effect of β-carotene supplementation on cancers in general and on lung cancers among smokers in particular. 22000 American doctors are participating in one of these supplementation studies (they take 50 mg β-carotene every other day), while 30000 Finnish smokers are taking part in another (they take 20 mg of β-carotene daily). These studies, which are sponsored by the National Cancer Institute in the United States, are based on a minimum daily intake of 15 mg of β-carotene, or 7 times the quantity in the average diet.

Treatment with vitamin A

About 50 million children in underdeveloped countries suffer from vitamin A deficiency which causes xerophthalmia, blindness and death. Although periodic massive dosing with vitamin A is not an ideal means of preventing vitamin A deficiency, WHO and UNICEF are running successful large-scale projects in Africa and South-East Asia, in which children are being given a weekly supplement of 100000 IU and mothers 200000 IU of vitamin A. The projects are being led by an ophthalmologist, Dr **Alfred Sommer** of the International Centre for Epidemiological and Preventive Ophthalmology in Baltimore, Maryland in the USA. Dr Sommer reports that xerophthalmia and blindness have both fallen by 80–90% (from 1.23 per cent to 0.15 per cent) while child mortality has fallen by 35%. These dramatic results underline the fact that vitamin A therapy can no longer be categorized as 'alternative' medicine, when the alternatives are blindness or death. The Task Force Sight and Life is gaining momentum in the fight to eradicate xerophthalmia in projects in Bangladesh, Bolivia, El Salvador, Ghana, India, Indonesia, Madagascar, Mexico, Nepal, Niger, The Philippines, Tanzania, Vietnam and Zimbabwe.

In Africa, vitamin A deficiency is the most common cause of measles-associated ulceration of the eye, which often causes blindness. Nearly one-third of the children with vitamin A-associated eye ulcers die during their hospital stay in Tanzania, despite vitamin A supplementation, report Drs **A. Foster** and **A. Sommer** in the *British Journal of Ophthalmology*.

Treatment for night-blindness involves a fortnightly supplement of 50000 IU, while treatment of dry and flaking skin can take the form of an ointment which contains 5000 IU per gram.

Vitamin A can assist the renewal of the skin cells and the collagen-proteins in the connective tissues. This is why vitamin A has long been employed in the treatment of a series of skin diseases.

Acne can be treated with a daily supplement of 100000–200000 IU over several weeks or months, although usually the high dosage is only prescribed for six weeks, after which time this is reduced to 50000–100000 IU. Alternatively, a two-week break in the high dosage can be held. Ointments containing vitamin A are also used in the treatment of acne.

Psoriasis has been treated with very high daily doses of up to 300000 IU, which can involve a risk of toxicity. New derivatives of vitamin A, isotretinoids, are therefore used instead in the treatment of psoriasis. These new vitamin A products have also come into use in the treatment of acne and other skin diseases and certain forms of cancer, such as T-cell leukaemia.

β-carotene (30–200 mg per day) is an effective and safe treatment for hypersensitivity to light and photo-allergy.

Vitamin A and β-carotene may be administered to cancer patients, particularly when they are undergoing cytostatic or radiotherapy. Dr **Clemens**, of the University of Tübingen in West Germany, has shown that the serum content of β-carotene drops significantly due to these therapies. This happens because these cancer therapies induce free oxygen radicals to kill cancer cells, but simultaneously they also consume the natural antioxidants of the body.

Professor **Leonida Santamaria**, of the Tumour Centre (Centro Tumori) at Pavia University in Italy, has shown that β-carotene supplementation (20–40 mg per day), together with another vitamin A derivative called canthaxanthine, has anti-cancer properties which, in addition, may well inhibit metastases (daughter tumours). Double-blind clinical studies are being carried out in Pavia in order to investigate these preliminary findings more closely.

Vitamin A poisoning

An excessive intake of vitamin A can produce symptoms of toxicity. The first examples of acute vitamin A poisoning were observed more than 100 years ago in polar explorers, who consumed a diet consisting of polar bear liver and seal meat, which contains very high concentrations (10 000–26 000 IU per gram) of vitamin A. Some of the explorers suffered from headaches, dizziness, nausea and vomiting, which are symptoms of acute vitamin A intoxication.

A daily intake which exceeds 5000 IU per kg bodyweight is too high and normally daily intake should not be more than 25 000 IU over an extended period, if this takes place without the supervision of a doctor. Early symptoms of over-consumption of vitamin A are itching and dry skin which can occur at a dosage of 20 000 IU per day. Symptoms such as fatigue and general discomfort can be an indication of disturbances in the central nervous system which are observed at long-term doses of 50 000–100 000 IU per day.

The treatment of vitamin A poisoning consists quite simply of discontinuance of intake. The brownish-yellowish skin coloration caused by β-carotene is not a sign of toxicity.

Vitamin A and pregnancy

Vitamin A is important for normal reproduction and for the growth and development of the foetus.

Pregnant women should, however, be aware that excessive intakes of vitamin A (retinol) can represent a threat to the foetus. This was discovered as far back as the 1950s, when experiments were conducted on animals in this field. 18 suspected cases of vitamin A-induced embryo damage have been detected in the United States, where the women concerned had consumed large doses (25 000–50 000 IU per day) during pregnancy. Dr **Barbara Underwood** and the Teratology Society in the United States, who have examined the circumstances of these cases, recommend an upper limit of 8000 IU per day during pregnancy and breast feeding. The new vitamin A derivatives (isotretinoids, e.g. Roaccutan) can also cause damage to the foetus. These new products are actually 30 times more concentrated than vitamin A itself.

Vitamin A and β-carotene in the blood

Normal plasma levels of vitamin A are 1.5–3.5 μmol l^{-1}. Values which are lower than this are either caused by insufficient intake or by a reduced capacity to absorb vitamin A in the intestines. Quantities of vitamin A in the tissues are dependent upon how much vitamin E is present, because vitamin A is easily oxidized if it is not protected by sufficient quantities of vitamin E.

β-carotene intake does not increase blood values of vitamin A (retinol). On the other hand, concentrations of β-carotene increase rapidly when we take food supplements or eat a lot of carrots. Professor **Fred Gey**, who is the head of WHO's vitamin project, recommends serum levels for β-carotene of 400 nanomoles per litre as a cancer prophylactic. Serum values for β-carotene are easily determined.

In a new study, which Dr **Tuomas Westermarck** and I carried out on 30 healthy volunteers, serum values of β-carotene were raised 3.5 times with a daily supplement of 6 mg, while a supplement of 12 mg per day raised β-carotene levels to almost 6 times the previous values. In my opinion, people who want to ensure adequate levels of β-carotene intake can suffice with a daily food supplement of 6–12 mg.

6. VITAMIN D

. . . and there is no new thing under the sun.

Ecclesiastes, chapter 1, verse 9

Human beings do not only get vitamin D from food, they also produce it themselves when the skin is exposed to sunlight. Vitamin D is a prerequisite for the absorption of calcium and phosphorus from the intestines and this is how vitamin D is needed for the mineralization of the bones. This is why vitamin D deficiencies in children cause rickets. Vitamin D is responsible for the regulating of blood coagulation and for the optimal functioning of the muscles and nerves. It is advisable for nursing mothers, infants and elderly people to take a vitamin D supplement, especially in the winter time. Overdosing can be dangerous.

In 1918 it was discovered that fish liver oils and sunshine had a curative effect on rickets. The latter effect was due to the production of vitamin D which occurs when sunlight reacts with the fats in the skin. Vitamin D in food is absorbed with fats in the digestive system, as is the case with vitamin A.

Quantities of vitamin D are also described in terms of international units (IU) and micrograms (μg). The RDA for adults is 400 IU, which is about 10 μg. Poor calcium absorption associated with vitamin D deficiency in adults can also lead to osteoporosis (brittle bones).

Vitamin D differs from other vitamins in that it is actually a related group of hormone-like substances (steroid alcohols). Vitamin D is not one single chemical substance. In fact, there are two different forms of vitamin D: vitamin D2 (ergocalciferol) and vitamin D3 (cholecalciferol). Both compounds are equally effective.

Vitamin D requirements and metabolism
Infants between the ages of two weeks and two years ought to receive 1000 IU per day, while those between the ages of 2 and 15 years should have a daily intake of 500–1000 IU. Vitamin D in food is a so-called provitamin because it does not actually become vitamin D until it is transformed under the influence of sunlight. It is therefore expedient to give children a supplement during the dark winter months.

In bright sunlight the skin produces up to 10 IU of vitamin D3 per square centimetre, depending on the colour (pigment) in the skin. Dark skin produces less vitamin D, since fewer of the ultraviolet rays reach the deeper layers of the skin where vitamin D is formed. Pollution of the ambient air also inhibits the penetration of the ultraviolet rays into the skin and thus decreases the ability of the skin to produce vitamin D. In general, however, we can forget all about vitamin D supplementation in sunny summer weather and on holidays in the sun!

Vitamin D requirements for older children and adults have not yet been set, because it is generally assumed that we produce enough from our contact with the sun. On the other hand, it is not a bad idea for elderly adults to take a supplement of 400 IU during the winter.

Provitamin D2 occurs naturally in sufficient quantities in vegetables and yeast, while provitamin D3 is to be found in fish oils, fatty fish (e.g. salmon, herring, mackerel, tuna fish and sardines), as well as in butter, cheese and liver. Milk, as such, is not a particularly good source of vitamin D! Commercial milk, including low-fat milk, is however generally fortified with vitamin D. In the USA, for instance, milk is fortified with 400 IU per litre, and in West Germany milk contains 750–1000 IU per litre. Vitamin D in foodstuffs is quite stable; storage, processing and cooking do not reduce vitamin D content.

Vitamin D from food is absorbed in the small intestine with the aid of the bile. Diseases in the liver, the pancreas, the intestines and the gall bladder can be instrumental in causing decreased vitamin D absorption. Disturbances in the absorption of fats as well as the use of anti-epilepsy medication can lead to vitamin D deficiency. D vitamins bind themselves to proteins in the blood. Vitamin D deficiency is diagnosed simply by analysing the concentration of 25-OH-vitamin D in serum.

Vitamin D — effects and treatment
Vitamin D functions like a hormone in the body, i.e. it has its effects far from the place where it is produced. It is essential to the organism's capacity to utilize calcium and phosphorus in the formation and strengthening of the teeth and bones. Vitamin D also regulates the permeability of the cell membranes. There is, however, a great deal which we do *not* know about the functioning of vitamin D in the organism.

Acute deficiencies are very rare. If there has been a deficit over an extended period, and this has reduced the absorption of calcium, it can give rise to osteoporosis. Vitamin D is therefore recommended as a prophylactic for this disease, which frequently attacks women after the menopause.

Previously it was generally believed that the vitamin D in breast milk would suffice for the newborn baby. This is obviously not the case. One litre of breast milk contains only 20–40 IU vitamin D, while the requirement of the infant is 400 IU per day. Therefore, supplements (400 IU per day) are also of importance to infants and children in the prevention of rickets, which has more or less been eradicated because of preventive measures. Adults occasionally need vitamin D supplements when the parathyroid glands fail to function, causing these glands' production of hormones to fall. Vitamin D deficiencies may cause another bone disease called osteomalacia.

Vitamin D can be consumed in the form of cod liver oil, cholecalciferol or ergocalciferol. An extra supplement of 200 IU (5 μg) per day is recommended during pregnancy.

Vitamin D poisoning

Whenever vitamin D deficiency is treated with supplements, it is advisable to follow metabolic changes like serum calcium and phosphate concentrations and the activity of alkaline phosphatase. When supplemented with vitamin D3, determinations of 25-OH-vitamin D3 suffices. Vitamin D is fat-soluble and is therefore stored in the body. If daily intake exceeds 50 000 IU there is an overt risk of overdose symptoms. Daily doses of over 25 000 IU over extended periods can lead to poisoning. Symptoms of this are unquenchable thirst, itchy eyes and skin, general discomfort, diarrhoea and frequent urination. A laboratory test will reveal increased calcium blood levels and X-rays will show accumulations of calcium in the blood vessels, the liver, the lungs and the kidneys. In general, I do not recommend daily intakes of more than 400 IU.

7.   VITAMIN E

Without knowledge life is only a shadow of death.

Moliére (1622–1673)

The active research on vitamin E has discovered that vitamin E, like selenium, has a number of important properties: both are antioxidants which apparently have anti-cancer properties and which help to increase resistance to viruses and bacteria. Vitamin E also protects against the undesirable side-effects of some cancer treatments, such as radiation treatment and chemotherapy. When vitamin E and the trace element selenium are taken together, possibly with other antioxidants, their cell-protective effects are multiplied many times.

Vitamin E cannot be produced in the human organism so we are totally dependent on the vitamin E in foodstuffs. The best sources are vegetable oils, whole wheat, broccoli, red cabbage and nuts. There are also small quantities in other fresh vegetables and eggs.

The food industry has to bear the responsibility for the loss of a great proportion of the vitamin E in foods during processing, deep freezing and preserving. The oxygen in the air can also destroy vitamin E in food. On the other hand, vitamin E or other synthetic antioxidants are added to most foodstuffs containing polyunsaturated fats, which have a tendency to become rancid quickly.

Ubiquinone or coenzyme Q10, which is also fat-soluble, resembles vitamin E in its molecular structure and in its function as an antioxidant in the cells' mitochondria.

Vitamin E requirements and metabolism

Vitamin E is a common name for different tocopherols (α-, β-, γ-tocopherol), known as potent fat-soluble antioxidants. Vitamin E content in foodstuffs and food supplements can be described in international units but it is more usual to use milligrams. 1 mg corresponds to 1 IU.

Actual requirements have not yet been established but the RDA is 10 mg for adults, slightly less for children and 3–4 mg for infants.

Many eminent researchers consider these recommendations to be much too low because they only take the prevention of actual deficiency symptoms into account. Vitamin E requirements are much higher when we want to utilize its beneficial antioxidant properties in the reinforcement of resistance, the prevention of cancer, cardiovascular diseases and other diseases.

Instead of the 'on the plate' recommendation of 10 mg per day, I would suggest blood tests in order to evaluate vitamin E concentrations. We now know that serum values of less than $18 \mu mol \, l^{-1}$ imply an increased cancer risk. It is therefore advisable to take as much vitamin E as is necessary in order to exceed this risk limit. My advice for prevention is 100–200 mg per day, or 10–20 times the RDA.

Severe deficiencies of vitamin E are rare and usually only affect newborn babies and people with reduced capacities for fat absorption. In animal experiments, on the other hand, vitamin E deficiencies have caused a wide spectrum of diseases and disturbances: sterility, increased risks of miscarriages, premature births, liver and kidney diseases, swelling of the prostate gland and muscular dystrophy. These deficiency diseases can be prevented and treated with vitamin E, selenium and other antioxidants.

Vitamin E blood levels in premature babies are often very low and it has been observed that these babies can develop symptoms similar to those of experimental animals which are deprived of vitamin E.

For the last 20 years we have known that vitamin E deficiencies in children can lead to disturbances in the functioning of the muscles and nerves and that these conditions can be cured with adjunct vitamin E therapy. Simultaneous deficiencies of vitamin E and selenium can be particularly dangerous in children, because this implies the loss of protection against free radicals.

In adults, vitamin E deficiencies can arise after major stomach surgery or diseases of the digestive organs. Common to these diseases is a reduced capacity for absorbing fat, and in such cases a corresponding deficiency of vitamin A will often occur.

Vitamin E deficiency can also result in a form of anaemia which cannot be cured with iron, but instead requires supplementation with vitamin E itself.

The effects of vitamin E

Antioxidant

It has long been known that vitamin E is the only chain-breaking lipid-soluble antioxidant which protects the cells from damage caused by free radicals and peroxides. The vitamin catches free radicals and destroys them before they can cause damage. This property is reinforced by the selenium enzyme, glutathione peroxidase.

If there is a simultaneous deficiency of vitamin E and selenium there is an even greater risk of rancidification (oxidation) in the cells. Vitamin E also inhibits the oxidation of vitamin A and therefore also protects against vitamin A deficiency, while vitamin C prolongs and reinforces the antioxidant effect. Vitamin E also protects essential fatty acids from oxidation (see Chapter 6).

Cancer

The potential cancer-preventive role of vitamin E has only been discovered very recently in connection with studies on selenium. It was first discovered that vitamin E reinforced selenium's capacity to prevent cancer in experimental animals. Supplementation with these two antioxidants reduced the incidence of cancer in rats by 50%. Epidemiological studies have later shown that vitamin E (and selenium) deficiencies are not uncommon in a 'healthy population' and that vitamin E may also be able to function as a cancer-prophylactic for human beings. The cancer risk-level would appear to be 18 μmol l^{-1} (about 7 mg l^{-1}). It therefore seems to make good sense to provide cancer patients with vitamin E supplements of about 200–400 mg per day.

Thrombosis and cardiovascular diseases

It would appear that prostaglandins, the so-called local hormones, are also connected with the functioning of vitamin E, although research in this area is still in the initial stages. We do, however, already know that high doses of vitamin E have an inhibiting influence on the aggregation of blood platelets (thrombocytes) and that it can help in the prevention of thromboses. When taken in megadoses, i.e. over 800 mg per day, vitamin E can operate as an anticoagulant.

On the other hand, it has been known for fifty years that vitamin E can help in the correction of cardiovascular disturbances. Many who suffer from the so-called window-watchers' disease (because they have to struggle just to be able to walk for short distances) have apparently benefited from vitamin E supplementation. The disease is called *claudicatio intermittens* and is caused by a circulatory disorder in the legs. Vitamin E therapy requires daily doses of 400–600 mg over extended periods (one to one and a half years) and if blood circulation improves, vitamin E levels must be maintained at a high level.

Doctors were already using vitamin E in the treatment of heart diseases in the 1940s, especially for coronary artery disease. The results in some individuals were remarkable but the early studies were not carried out with control groups. Later controlled studies on the same area gave more disappointing results. If vitamin E is applied as the only treatment for heart patients, the results are statistically no more impressive than treatment with placebo medicine.

On the other hand, more is expected of studies where vitamin E is administered in tandem with other, water-soluble antioxidants, particularly selenium and a completely new water-soluble analogue of vitamin E. Animal experiments have provided more than just a hint of a protective effect on the heart, although these have involved very high dosages, compared with the RDAs for these antioxidants. Studies conducted by Professor **Kenth Ingold**, of the National Research Council of Canada, indicate that a combination of fat-soluble (vitamin E) and water-soluble antioxidants produce the best response in protecting the heart.

Dr **Jan Häggendahl**'s research team has recently shown that an injection of vitamin E and selenium protected the heart muscles of pigs, weighing 70–90 kg, against the consequences of experimentally induced myocardial infarction of the heart. For many years, I have been recommending a daily supplement of 400 mg of vitamin E plus other water-soluble antioxidants to my heart patients.

The immune system
It would appear that vitamin E can improve the activity of the immune system and thereby strengthen resistance to a range of diseases, including viral and bacterial infections. Experiments with animals have shown that in order to achieve this effect a dosage of 5–20 mg/kg per bodyweight is required, corresponding to 350–1400 mg per day for a person weighing 70 kg.

Antidote to toxic substances and cytostatics
Vitamin E may also have a detoxifying effect in connection with heavy metals and reduce the harmful effects of cytostatics like adriamycin. Vitamin E also has a protective effect against overdoses of selenium and the heart medicine, digoxin.

Cataracts
Large-scale population studies in the USA and Canada have shown that people who take 300–400 mg of vitamin E per day very rarely get cataracts. β-carotene and vitamin C reinforce this property of vitamin E.

Parkinson's disease
New discoveries in Canada and the United States suggest that Parkinson's disease is at least partially caused by the activities of free radicals, implying that adjunct antioxidant therapy may be beneficial for this illness. Since 1979, **Dr Stanley Fahn**, Head of Parkinson and Movement Disorder Research at the Columbia University College of Physicians and Surgeons, has been treating Parkinson's disease patients with antioxidants, in the form of 400–3200 mg of vitamin E and 3000 mg of vitamin C per day. Dr Fahn, speaking at a conference on vitamin E in New York in October 1988, said that his results showed that these vitamins could slow down the progression of Parkinson's disease. His experiences in this field have also been published in *Archives of Neurology* (July 1988), the neurologists' own journal. Dr Fahn's results showed, moreover, that treatment with a medicine called levo-dopa, or other similar traditional anti-Parkinson's medicines, could be postponed for several years because the progression of the disease could be delayed with antioxidant vitamin therapy. These relatively large daily doses of vitamins C and E were well tolerated and no signs of toxicity have been observed.

Vitamin E therapy
People who consume large quantities of polyunsaturated fats from vegetable oils or plant margarine need extra vitamin E, in the region of 100–200 mg per day, in order to protect against the rancidification of fats during cell metabolism (lipid peroxidation).

I recommend doses of 600 mg per day for sufferers of *claudicatio intermittens* but it often takes from several months to one and a half years before possible beneficial effects appear. I also recommend adjunct therapy consisting of 400 mg of vitamin E, plus selenium and other antioxidant vitamins and minerals, to heart patients suffering from *angina pectoris* or those who have recently suffered a heart attack.

Vitamin E requirements can sometimes be as high as 1000 mg per day and the purpose of these large doses is to prevent new blood clots or to improve the sensitivity of the heart muscle to oxygen deficiency, or, for that matter, to inhibit the advancement of cataracts or Parkinson's disease.

It is not inappropriate to try vitamin E therapy in connection with pregnancy, breast feeding, impending miscarriages, or in the event of childlessness.

Attempts have also been made to use vitamin E therapy in the treatment of a number of muscle diseases and for muscular dystrophy but the results have not been particularly convincing.

I usually prescribe large doses of vitamin E, selenium and other antioxidants for cancer patients — not as the only treatment but as adjunct therapy.

Studies published by Professor **Volkmar Böhlau** in West Germany show that vitamin E can improve the general health of the elderly. My own team has observed a significant improvement in the psychological and physical health of elderly people when they were supplemented with a daily dose of 400 mg vitamin E and selenium (see 5.3).

Overdoses of vitamin E

Like the other fat-soluble vitamins, vitamin E is stored in the body, but, despite this, there is no risk of overdosing. No cases of poisoning have ever been registered in connection with the taking of capsules or tablets of vitamin E. Canadian studies in the 1940s involved doses of up to 10 g (1000 times the RDA) over extended periods, without any signs of poisoning. Megadoses (over 1000 mg per day) can, however, lead to stomach problems and diarrhoea for people with a delicate stomach. People on anticoagulant therapy should not take vitamin E in daily doses exceeding 800 mg.

8.   VITAMIN K

All knowledge is of itself of some value. There is nothing so minute or inconsiderable that I would not rather know it than not.
Samuel Johnson (1709–1784)

Vitamin K is essential for the coagulation of the blood. It occurs naturally in plants and in animal livers. In human beings, vitamin K is produced by the intestinal bacteria and deficiencies of this vitamin are therefore very rare. Serious deficiencies can nevertheless occur in newborn babies, in older children and in adults with abdominal and intestinal diseases, e.g. celiac sprue. Vitamin K would appear to play an important role in the prevention of osteoporosis.

Vitamin K was discovered by the Danish Nobel Prizewinner, **Henrik Dam**, in 1935. He was working at Copenhagen University when he discovered that a mixture of putrefying fish powder and alfalfa could prevent haemorrhaging in chickens which otherwise had a tendency to haemorrhaging because of their low-fat diet. Henrik Dam isolated a compound from alfalfa, which he called 'koagulation vitamin' (coagulation is spelt with a K in Danish). The K has been retained in all languages as the term for this vitamin.

There are two naturally occurring forms of vitamin K: vitamin K1 (chemical name phytomenadione) occurs in the leaves and other parts of green plants, while vitamin K2 (menaquinone) is formed by intestinal bacteria. There are also synthetic forms, vitamin K3 (menadione), K4 (menadiole) and K5 which have a biological effect which is twice as strong as that of natural vitamin K.

Natural forms of vitamin K are yellowish oils which are resistant to heat, the effects of air and damp but which decompose on contact with light. Vitamin K is quickly destroyed when it is exposed to ultraviolet light or to the sun's rays.

Vitamin K requirements and metabolism
There is no RDA for vitamin K but it is estimated that human beings need at least one milligram of vitamin K per day. Deficiencies can occur in connection with alcoholism and in disturbances in the absorption of fats or in the event of liver and intestinal diseases, e.g. coeliacia. Vitamin K deficiencies are likely to occur in patients receiving antibiotics which suppress normal gastrointestinal flora, e.g. in patients receiving treatment for burns. The patient is then exposed to the risk of internal haemorrhaging because the blood is not able to coagulate.

In addition to antibiotics, pain-killers containing acetylsalicylates can also produce an increased requirement for vitamin K. It is also normal practice to give injections of vitamin K before various different forms of surgery, especially gall bladder operations. The interference of large doses of vitamin E (over 800 mg per day) with vitamin K activity has recently been confirmed.

Newborn babies have no intestinal bacteria — the intestines are first populated by bacteria one or two days after birth and this is why newborn babies are often given an injection of vitamin K to avoid haemorrhaging.

Animal products contain very small quantities of vitamin K, although small amounts do occur in liver, meat, cod liver oil and egg white. Plants, on the other hand, are rich in this vitamin. Spinach, cauliflower, peas, soya beans and carrots are good sources. Tomatoes, grains and honey are also sources but they are of less significance. The average diet yields 0.3–0.5 mg of vitamin K. Like other fat-soluble vitamins, vitamin K is absorbed together with fats in the intestines. It is transported into the bloodstream via the lymph system.

No actual diseases which are caused by vitamin K deficiency alone have been discovered in human beings.

The effects of vitamin K
The only traditionally known effect of vitamin K is its role in building up the capacity of blood to coagulate. It participates in a number of coagulation factors, including prothrombin. The anticoagulant medicine, dicoumarol, resembles vitamin K in its structure and counteracts the effect of vitamin K. Megadoses of vitamin E may also inhibit blood coagulation, and in such cases vitamin K supplements may be necessary.

Vitamin K1 (phytomenadione) deficiency may be a risk factor for osteoporosis (brittle bones). It has been known for a decade that vitamin K1 has a crucial role in normal bone formation. It catalyzes the metabolism of osteocalcin, which is the protein matrix for new bone formation. With the help of vitamin K1 osteocalcin binds calcium ions leading to normal bone calcification. Without sufficient vitamin K1, normal bone calcification presumably does not proceed. In the United States, Dr **Jonathan V. Wright**, who has studied the role of vitamin K1 in osteoporosis, suggests the testing of serum vitamin K1 as a useful preventive and therapeutic tool for osteoporosis. Women with existing osteoporosis would appear to have much less chance of recalcification if suffering from vitamin K1 deficiency.

Vitamin K1 deficiency is more likely among those who do not eat green vegetables and those whose digestion and assimilation of fats and fat-soluble vitamins is impaired. Dr Wright recommends serum K1 testing for everyone with osteoporosis and for women whose related elders have osteoporosis. Dietary changes or supplementation, or both, would be required for those with low or low-normal levels of serum vitamin K1. Repeat testing almost always shows substantial improvements in just a few weeks, according to Dr Wright. Vitamin K1 testing may not be valid for persons who are on anticoagulant therapy.

Studies indicate that vitamin K supplementation of osteoporotic women may reduce the calcium loss from their bones by 18–50%, according to a Japanese study by Dr **A. Tomita**. Vitamin K supplementation has also been shown to accelerate healing of experimental fractures in rabbits.

Water-soluble forms of vitamin K, such as menadiol sodium diphosphate (available in 5 mg tablets) or phytomenadione (10 mg chewing tablets or injections), may be administered to prevent and to treat vitamin K deficiency.

Treatment with vitamin K

Most newborn babies are given an injection of vitamin K, while long-term diarrhoea in infants can require further supplementation with vitamin K.

Occasionally, supplements of vitamin K are also given to people with alcohol-damaged livers or in connection with long-term treatments with antibiotics, e.g. for serious and extensive burns.

The need for supplements of vitamin K can be determined by investigating the time taken for the blood to coagulate.

Vitamin K tablets can be bought over the counter at the chemist's, but should never be taken without first consulting a doctor.

Adults can tolerate doses in the region of 20–40 times the normal daily dose, but such large doses can give rise to minor discomfort, nausea or vomiting. In addition, it is advisable to be cautious with large doses for patients with liver diseases, because this can result in the direct opposite of the desired effect.

9.   VITAMIN C

A snake lingers in the tree of knowledge. Whosoever approaches it will be an unhappy victim of doubt for all eternity.

Lauri Viita (1916–1965)

Up to a fifth of the population of Scandinavia suffers from a serious deficiency of vitamin C in the spring. In June the new potatoes and fresh fruit arrive just in time to

prevent scurvy. These new research results show that vitamin C deficiency still represents a serious problem for health in Scandinavia.

Vitamin C has many important functions in the organism. It increases the absorption of iron, it contributes to the formation of the bones, the teeth and the tissues, it speeds up the healing of wounds, it helps to maintain the elasticity of the skin, it is needed for the production of stress hormones and it improves the resistance to infections, high blood pressure, arteriosclerosis and cancer.

Vitamin C was synthesized for the first time in 1933 and has since been the object of intensive research. Nevertheless, the actual significance of vitamin C is still a topic of lively discussion amongst researchers.

All animals need vitamin C, but apes, guinea pigs, certain species of bats and human beings cannot produce the vitamin themselves and therefore require supplies of it in their diet. The best sources of vitamin C are berries, fruits, vegetables and potatoes. A great deal of vitamin C is lost when the foodstuff is heated, deep-frozen or preserved. Vitamin C is also sensitive to light.

A whole series of lesser known fruits and plants have a vitamin C content of more than 1000mg/100 g of the flesh of the fruit. Rosehips, for example, contain 1250 mg/100 g, while broccoli, cauliflower, green mango fruit and kiwi fruits are very good sources of vitamin C with 100 mg/100 g. Citrus fruits on the other hand, which are otherwise generally considered to be the best sources of vitamin C, contain 'only' 50 mg/100 g fruit flesh. Spanish peppers and many tropical fruits contain five times as much vitamin C as citrus fruits.

Vitamin C requirements and metabolism

The RDA for vitamin C is 60 mg for healthy adults. Vitamin C, which can be measured in the blood, is stored in the body, enabling us to draw on our reserves for several months.

Just 10 mg per day is enough to cure and prevent scurvy, but it is not enough to build up vitamin C reserves in the organism. If an adult takes 60–75 mg on a daily basis, about 1500 mg will eventually be stored in the body, which is sufficient to prevent scurvy for up to three months if vitamin C is then excluded from the diet.

Many people receive only very small quantities of vitamin C during the winter months, while the risk of vitamin C deficiency is higher for slimmers, smokers and the elderly.

A number of factors increase vitamin C requirements. This applies to very warm and very cold environments, smoking and contraceptive pills. Measurements of the blood plasma of smokers show that their vitamin C levels are 20–40% lower than those of non-smokers.

Certain of the larger animals, dogs for example, are able to produce vitamin C themselves. Production corresponds to 10 mg per day per kg bodyweight. Apes, on the other hand, are unable to produce vitamin C themselves, and, according to laboratory experiments, they need 55 mg per kg. This corresponds to 3.8 g for an adult human. The question therefore arises as to what one considers to be a sufficient intake, when one sets the level for an RDA (Recommended Dietary Allowance). 60 mg of vitamin C is certainly enough to prevent scurvy, but it is doubtful whether this is enough to protect the brain from ischaemia (oxygen deficiency) or to optimize the functioning of the immune system.

The existing recommendations are purely concerned with prevention of the classical deficiency symptoms and take no account of the recently discovered properties of vitamins and minerals. For the improvement of resistance and the prevention of cancer and cardiovascular diseases quite different and larger doses may be required.

The effects of vitamin C

Vitamin C is an important antioxidant for the cells. In addition, it has a reinforcing influence on the effects and the period of activity of other antioxidants, such as vitamins A and E. These three vitamins have, on the whole, a synergistic relationship, i.e. they reinforce each other.

Vitamin C would appear to play a particularly important role in the brain tissues because it protects against ischaemia (oxygen deficiency). Ischaemia and the subsequent reperfusion (see p. 51–52) increase the production of oxygen free radicals which may lead to cell destruction. Moreover, nerve tissue is the part of the body in which the highest concentrations of vitamin C are found, and the part of the body where this vitamin is most required.

Cancer prevention

Animal experiments have revealed that vitamin C, like selenium and vitamin A, may prevent the occurrence and development of cancer. Vitamin C may also prevent a range of substances from developing into carcinogens. Large quantities of vitamin C can prevent nitrites and nitrates, which are virtually impossible to avoid in our foodstuffs, from being transformed into *nitrosamines*, which can otherwise lead to the development of cancers in the stomach and the intestinal tract.

Population studies have also indicated that a high intake of vitamin C can reduce the risk of contracting cancers of the stomach and the oesophagus, presumably as a result of the inhibition in the production of nitrosamines. This preventive effect may also extend to other forms of cancer, including cancers of the throat and womb and malignant melanomas.

The formation of connective tissue

Vitamin C is also a prerequisite for the production of the connective tissue protein, collagen. Vitamin C is therefore necessary to ensure a more rapid healing of wounds. Vitamin C also helps the skin to maintain its youthfulness and elasticity.

Vitamin C is an important factor in the formation of the bones, and is therefore vital to the healing of broken bones.

Protection for the heart and blood vessels

Vitamin C requirements for the heart muscle are greatly increased in the event of hardening of the coronary arteries or after a heart attack. It has been discovered that the white blood corpuscles transport large quantities of vitamin C to the heart from other parts of the body. The consequence of this is that vitamin C levels in other tissues and in the serum can fall if the patient is not given vitamin C supplementation.

Vitamin C contributes to the prevention and repairing of cell damage in the blood vessels of the heart muscle. Intravenous injections of vitamin C can reduce the tendency of blood platelets to aggregate, thus reducing the risk of blood clots and arteriosclerosis.

Population studies (in Finland, for example) have indicated that there is an inversely proportional relationship between vitamin C intake and high blood pressure and arteriosclerosis. Other factors, such as the level of dietary fibre intake, also play a role and it is therefore difficult to judge the degree to which each individual factor contributes to the prevention of these diseases.

Vitamin C may also have an influence in reducing LDL-cholesterol levels (the 'dangerous' cholesterol) and in increasing concentrations of HDL-cholesterol (the 'good' cholesterol). In addition, antioxidants, such as vitamin C, appear to inhibit LDL-cholesterol from oxidizing into the dangerous o-cholesterol (oxidized cholesterol), which forces its way into the artery walls and contributes to the development of arteriosclerosis. This effect was first revealed in a large-scale study conducted by Professor **Fred Gey**, a researcher with WHO, and is seen particularly in people with both a high level of cholesterol and a low intake of vitamin C. On the other hand, it does not appear that vitamin C further reduces normal cholesterol levels.

In China, megadoses of vitamin C, in the form of injections, are successfully administered to patients suffering from a specific heart disease, a cardiomyopathy called Keshan's disease (see also 5.3 on selenium).

Strengthening our powers of resistance
Vitamin C improves phagocyte function. Phagocytes are white blood corpuscles whose task is the detection, destruction and consumption of virus, bacteria and cancer cells. Human beings with a vitamin C intake of 2–3 g per day have an increased activity of both phagocytes and neutrophile white blood corpuscles.

One gram of vitamin C per day reinforces the immune system by increasing the blood content of antibodies. It may be that this occurs simultaneously with an increase in the production of interferon.

Furthermore, vitamin C can be an aid to the prevention and self-treatment of allergies, because the vitamin inhibits the deleterious effects of histamine. Histamine is released in association with allergic responses in the skin, the respiratory passages and in the mucous membranes in the nose. The consequences of histamine release are runny nose, breathing difficulties, itching and nettle rash. This sort of histamine release can also be precipitated both by warm or cold air.

Vitamin C as an antidote
Vitamin B1, vitamin C and the sulphur-containing amino acid, cysteine, can prevent formalin, formaldehyde and acetaldehyde from causing damage to the organism. Tobacco smoke, for example, produces formaldehyde, while acetaldehyde arises in connection with the combustion of alcohol in the body. The red colour in the skin which is associated with alcohol consumption is caused by acetaldehyde.

Vitamin C also counteracts other toxic and carcinogenic effects of nicotine, tobacco smoke, vehicle exhausts, nitrogen compounds and the heavy metal, cadmium. It is precisely because the organism uses up vitamin C as an antidote to these various toxic substances that smokers and alcoholics have serum values of this vitamin which are 20–40% lower than other people's. This deficit can be compensated for by sizeable supplements of vitamin C and it may be that this — together with selenium, zinc, β-carotene, ubiquinone and vitamins A and E — can contribute to the mitigation of some of the harmful effects of smoking.

Overdosing with vitamin C

Normally it is not possible to overdose with vitamin C, but when very large doses are involved, i.e. several grams per day, this can lead to stomach trouble, especially for people who already have a delicate stomach or who have a tendency towards stomach ulcers or gastric catarrh. Stomach problems can usually be avoided by taking vitamin C in the form of a soluble or time-release tablet or after meals, or, if necessary, by taking medicines which neutralize gastric acids.

Sudden large doses of vitamin C can cause diarrhoea but this is easily avoided by gradually working up to large doses. On the other hand, a sudden cessation in vitamin C intake can lead to fatigue. The daily dose should therefore be reduced just as gradually as it is increased, if one no longer feels that such large doses are necessary.

Vitamin C can, in some rare cases, precipitate an allergic reaction, caused by ascorbic acid.

Oxidation of vitamin C

Vitamin C is an antioxidant and can therefore itself be oxidized very easily. Vitamin C itself is not one of the strongest of the antioxidants, but in the tissues it reinforces the effects of other, fat-soluble, antioxidants.

When vitamin C is exposed to air, light or heat a brownish compound is formed which gives off a burnt smell. This is dehydroascorbate (DHA) which has a diametrically opposed effect from vitamin C.

DHA is a pro-oxidant, i.e. the opposite of an antioxidant, and it stimulates the production of oxygen free radicals and leads to cell damage. It is not inconceivable that some of the negative results from animal experiments on vitamin C have been caused by the formation of DHA in the animals' feed. Half of a given quantity of vitamin C dissolved in water becomes DHA within 2–4 hours.

The iron and copper in food also speed up the oxidation of vitamin C, although the cell membranes in the plants protect it from destruction. However, the cell membranes are dissolved by cooking, rapidly causing the vitamin C to be wasted or to be transformed into DHA. Vitamin C in tablet form retains its effectiveness for longer than the vitamin C in food.

10.   THE B VITAMINS

He who lacks the ability to perform a certain task is always capable of something else.

Kenyan proverb

B vitamins are most commonly used in the treatment of nervous problems, fatigue, stress and in the prevention of alcohol problems. The B vitamins participate in all the most important aspects of food metabolism and in the production of energy. Children

require B vitamins for normal development and these vitamins are also essential to the renewal of the cells. Some of the B vitamins would also appear to contribute to the reinforcement of the immune system and to the prevention of blood clots. In addition, most of the B vitamins are also antioxidants.

The B vitamins are a group of substances, some of which are not actual vitamins. We generally get enough B vitamins from our diet to avoid deficiency symptoms. Those who drink too much alcohol, women who take contraceptive pills, the elderly and children on antibiotics are particularly susceptible to deficiencies of B vitamins.

The group of B vitamins is composed of eight actual vitamins and at least five vitamin-like compounds. All of these occur in a normal mixed diet. Especially good sources of B vitamins are yeast, liver and wheat bran. All of the B vitamins are so-called coenzymes and they are classified according to their various functions, whether this be energy production, blood production or other functions (see table, p. 125).

The B vitamins therefore have a manifold influence on our health. They are required in the functioning of the immune system, the digestive system, the heart and the muscles and in the production of new blood cells.

It has only recently been discovered that several of the B vitamins have antioxidant effects and that they stimulate the activity of the immune system. In addition, some of the B vitamins have a number of other properties which were unknown only a few years ago. Vitamin B6, for example, helps to prevent the platelets from aggregating. This vitamin also appears to have a beneficial effect on herpes, depression and, in some cases, menstrual discomfort.

Each of the B vitamins has its own special function in the metabolic processes and in this cannot be replaced by any of the other vitamins in the group. Nevertheless, they work closely together and deficiencies of a single B vitamin are rare. On the contrary, a deficiency of one of the B vitamins usually leads to deficiencies of several of the others.

Vitamin B deficiency is rarely expressed as a disorder in a specific organ. It is more likely to be accompanied by a whole series of symptoms, not all of which automatically lead one to think of B vitamin deficiency. Only the most severe deficiency conditions result in the classical deficiency symptoms of pellagra and beriberi, which are now extremely rare diseases in the western world.

On the other hand, it is not nearly so uncommon to see symptoms of a more minor and concealed deficiency condition. These arise as a consequence of the widespread use of refined foodstuffs, bread made from refined white flour, increased sugar consumption, smoking, alcohol, contraceptive pills and antibiotics.

The relationship between the individual B vitamins can be further demonstrated by the fact that an increased supplement of one of the B vitamins can increase the requirements for the others. This is why vitamin B therapy is usually administered in the form of a complex of the different vitamins in the group, to patients with rheumatoid arthritis, for example. There are a number of different products which contain most of the B vitamins, such as yeast tablets.

Some doctors prescribe much larger doses of vitamin B6, compared to the RDA, in the event of hardening of the coronary arteries, herpes infections or menstrual problems.

Riboflavin (B2) can be used as the sole treatment for hacks in the corner of the mouth, while vitamin B12 and folic acid are prescribed for the treatment of deficiency symptoms. New research results indicate that these two vitamins also may inhibit the spreading of lung cancer (metastases).

The dietary B vitamins are water-soluble and are therefore not stored in the body. This means that we are entirely dependent upon a daily supply from our diet. It is very unusual to hear of overdosing with the B vitamins, because they are water-soluble, although very large doses over extended periods can cause headaches, numbness, sweating and insomnia. These symptoms disappear very quickly if intake is reduced.

In the USA and Australia there have recently been a number of cases of vitamin B6 poisoning, but these cases involved megadoses of up to 1000–7000 mg per day, which corresponds to 500–2500 times the RDA.

Not all B vitamins have a number and in recent years it has become increasingly normal to use the chemical term, also for B vitamins which already have a number. In

---

Division of B vitamins according to function

Energy-producing
1. Vitamin B1 (thiamin)
2. Vitamin B2 (riboflavin, lactoflavin)
3. Niacin, nicotinate, nicotinamide
4. Pantothenate
5. Biotin

Blood-producing and cancer-inhibiting
1. Folic acid (folate, folacin)
2. Vitamin B12 (cyanocobalamin)

Other functions
1. Vitamin B6 (pyridoxine)
2. Choline*
3. Inositol (myoinositol)*
4. Para-aminobenzoic acid (PABA)*
5. Ubiquinone (coenzyme Q)*
6. Lipoic acid*

---

*Vitamin-like substance (not an actual vitamin)

---

the following sections on the functions of the individual B vitamins I will therefore mainly use the chemical name instead of B1, B2, B3, etc. Very recently, the pharmaceutical industry has been able to synthesize many of the water-soluble B vitamins.

11.   THIAMIN (VITAMIN B1)

I love reality. It gives the bread its taste.

Jean Anouilh

Thiamin is a water-soluble vitamin which is highly sensitive to being heated up. It

occurs in considerable quantities in yeast, grain husks, whole-wheat products, peas, soya beans, peanuts, eggs, pork and liver. Thiamin is responsible for energy production in the cells, it improves the functioning of the T-lymphocytes (a component of the immune system), and it is essential to the normal development and functioning of the brain, the muscles and the nerves. In addition, thiamin is an antioxidant.

The preparation of food often leads to the loss of as much as 30% of thiamin content and this can result in a thiamin intake which is lower than the recommended levels. In Finland thiamin intake is lower than normal in 10% of men and 5% of women.

A healthy adult needs 0.5 mg of thiamin per 1000 calories, which corresponds to about one milligram per day. The RDA for thiamin is 1.5 mg. Thiamin requirements are closely associated with the amount of food which is consumed. Pregnant women need more, and it would appear that the elderly also have a higher daily thiamin requirement, because the activity of this vitamin decreases with age.

Operations, stress and various diseases also increase thiamin requirements. Moreover, thiamin needs are also increased by smoking, high alcohol intake and a high proportion of carbohydrates in diet.

Thiamin deficiency is common in countries where rice is the staple diet, while in the industrialized world it is estimated that about half of the population suffers from long-term, although minor, thiamin deficiency.

In many countries, diets are so poor in thiamin that if the quantities of intake are transposed to animals, i.e. if corresponding quantities per kg bodyweight are given to experimental animals, this causes deficiency symptoms such as irritability, depression and anxiety. It is not inconceivable that these symptoms in human beings may also be connected to some extent to thiamin deficiency. The individual human requirement for this vitamin varies because of different levels of alcohol, tobacco, medicine and carbohydrate consumption. Alcoholics often suffer from thiamin deficiency, which can lead to a weakening of the heart muscle and even heart failure.

Thiamin deficiency may arise in connection with disturbances in nutrient absorption in the intestines, or with anorexia nervosa, or with long-term poor general health associated with diarrhoea and vomiting. Diuretic medicines increase the elimination of this vitamin with the urine.

There are many factors which contribute to the destruction of this vitamin. Grains contain a lot of thiamin but this is often lost in the process of the refinement of grains, while the rest can disappear through baking, cooking or frying. Thiamin is also destroyed by the oxygen in the air and by water.

Thiamin and thiamin salts are absorbed in the small intestine, but a proportion of the thiamin which is taken as food supplementation tablets is not absorbed at all. In addition, some of the thiamin which has already been absorbed in the bloodstream may return to the intestine with the bile, in which case it is transported out of the body with the excreta.

As with the other water-soluble vitamins, thiamin cannot be stored in the organism in large quantities. An ample intake of thiamin rapidly fills up the body's modest reserves and the excess is eliminated with the urine, which can have a special smell, if the elimination of the vitamin is abundant. The greatest concentrations of thiamin are to be found in the liver, the heart and the brain.

A proportion of the thiamin in the organism is not utilized for its genuine function because it is 'consumed' by certain foodstuffs and stimulants, such as coffee, some teas, alcohol, tobacco and hormone products like contraceptive pills.

The effects of thiamin
Thiamin promotes growth, appetite, digestion and the functioning of the intestinal and nervous systems. It has also been maintained that thiamin is beneficial for mental health as it participates in the functioning of the nerves, the muscles and the heart. Thiamin is also reputed to be beneficial against sea-sickness, toothache and other minor pains. Thiamin is also a mild diuretic.

Even minor deficiencies of this vitamin can cause disturbances in the peripheral nervous system and lead to sleeplessness, loss of appetite, fatigue, depression, irritability, failing memory and loss of concentration. Symptoms of thiamin deficiency can, for example, occur in connection with pregnancy, but these symptoms are easily inhibited by food supplementation with this vitamin.

Thiamin functions as an antioxidant in the organism and it participates in the coenzyme which is responsible for the metabolism of carbohydrates to fat, which in turn produce the energy for the cells. Other B vitamins which also participate as coenzymes in this process are pantothenate, nicotinate and riboflavin.

Together with the other antioxidants, thiamin inhibits the production of free radicals and thus protects the cells from acetaldehyde, a substance which can cause mutations and cancer.

Thiamin is not particularly toxic even in large doses. Even very high doses of up to 500 mg (more than 300 times the RDA) have been used over extended periods without any evident harmful effects. If thiamin is administered in the form of an injection, it can give rise to symptoms of toxicity, such as swelling, shaking, nervousness, disturbances in heart rhythms and allergic reactions. Extremely large injections can lead to disturbances in heart and nerve functions. The treatment for symptoms of overdosing involves quite simply cutting down or stopping intake altogether.

Thiamin therapy
Pregnant women and breast-feeding mothers ought to take a moderate supplement of thiamin, and this also applies to women who take contraceptive pills. Thiamin supplementation can also be necessary in stressful situations and in long-term illnesses, because both of these situations increase thiamin requirements drastically.

Thiamin can alleviate a smoker's cough and can be used in the adjunct therapy of a range of illnesses which increase the production of free radicals and weaken the immune system. These incorporate skin diseases, arteriosclerosis, disturbances in heart function, arthritis and related disorders, cancer and many others.

Diabetics who take insulin should be wary of taking supplements of vitamin B1 together with vitamin C and cysteine, because this combination of substances can render insulin ineffective. In addition, it is common practice to treat polyneuropathy, a disturbance of the peripheral nervous system which is associated with diabetes and which affects the sense of touch, with thiamin and vitamin B12. As yet, however, these effects have not been confirmed in controlled studies.

All the B vitamins are synergistic, i.e. they reinforce each other's effects. This is why it is advisable to take thiamin as part of a vitamin B complex.

12.   RIBOFLAVIN (VITAMIN B2)

First get hold of the facts, and then you can distort them as much as you like.
                                                        Mark Twain (1835–1910)

Riboflavin (lactoflavin) is a yellow, water-soluble colorant, which is partially decomposed by sunlight and heat. There is riboflavin in all of our cells, where it produces energy and increases the formation of antibodies. Riboflavin deficiency causes hacks in the corners of the mouth and oversensitivity to light. There is a great deal of riboflavin in the eye, but it has not yet been ascertained whether or not this vitamin is of special significance for sight.

Riboflavin is every bit as essential as oxygen. In fact, one of the functions of riboflavin is to ensure that oxygen is used in the production of energy. In addition, it regulates the metabolism of carbohydrates, proteins and fats. Deficiencies of this vitamin lead to comprehensive disturbances in the metabolic processes. Despite this, no actual diseases are known which are caused by riboflavin deficiency.

According to one textbook, *Nutritional Support and Medical Practice* (1983), riboflavin deficiency is the commonest vitamin deficiency condition in the industrialized countries. According to one study, conducted in England in 1977, one hospital patient in five suffers from a deficiency of this vitamin.

Riboflavin was isolated from milk for the first time in 1933 and was synthesized two years later. The substance is yellowish orange in colour and crystalline.

Riboflavin requirements and metabolism

Requirements of riboflavin are slightly different for men and women. The requirement for women is 1.1–1.3 mg per day, while for men it is slightly higher and for children slightly lower. Infants can suffice with an intake of 0.4–0.6 mg.

Deficiencies of riboflavin are most common amongst children, pregnant women and nursing mothers, slimmers and the elderly. Deficiency symptoms can sometimes be observed in vegans, because their diets do not contain milk products. Various forms of tranquilizers, contraceptive pills, antibiotics and sulpha products can also lead to riboflavin deficiencies.

Another group who are prone to deficiencies of riboflavin are the chronically ill. Tuberculosis, arthritis, heart failure, major surgery and long-term treatment for burns are examples of diseases or situations which increase the risk of deficiencies.

Good sources of riboflavin are milk, cheese, eggs, yeast, liver, kidneys, fish and vegetables.

Riboflavin absorption takes place in the small intestine and this process is regulated by hormones from the thyroid gland. Some nerve-medicines and antidepressants can give rise to disturbances in absorption. About 10% of daily intake is eliminated with the urine and a smaller percentage with the excreta. Some of the riboflavin which is eliminated with the excreta does not stem from intake, but from intestinal bacteria, which produce riboflavin themselves. It would appear, however, that the organism is unable to absorb this bacteria-produced riboflavin.

A tough slimming programme or diabetes increases the elimination of riboflavin in connection with the breakdown of proteins.

Riboflavin reserves in the organism seem to be very stable. Even in a deficiency period the reserves only drop by about 30–50%. On the other hand, if diets are very low in protein, this draws on reserves to a greater extent. The greater part of our riboflavin is located in the red blood corpuscles. Direct measurements of blood levels of riboflavin do not, however, give the most reliable information about riboflavin values in the organism. A more useful method is to measure the quantities of the vitamin which are eliminated with the urine, as this falls when there is an insufficient intake.

The effects of riboflavin

Riboflavin is of importance for the production of essential fatty acids, niacin, noradrenaline, serotonin, histamine and acetylcholine. It also participates in a number of enzymes which are part of the antioxidant system.

Riboflavin also inhibits the production of prostaglandins and leukotrienes. These substances are hormone-like substances which stimulate inflammation reactions and contractions of the respiratory passages. Riboflavin deficiencies in experimental animals have furthermore resulted in embryo damage, which resembles that caused by thalidomide. Conversely, attempts to induce embryo damage chemically in experimental animals have been prevented with riboflavin supplementation. No information is available, however, about the extent to which these results might apply to human beings.

Severe deficiencies of riboflavin are rarely seen in human beings, but symptoms of marginal deficiencies are hacks in the corners of the mouth, itchy tongue, infections in the oral cavity, and changes in the skin around the nose. This deficiency can also lead to seborrhoic eczema, i.e. a flaking and greasy rash.

Occasionally, deficiencies express themselves in the form of oversensitivity to light and prominent veins in the white of the eye. All these symptoms do not necessarily indicate riboflavin deficiency, but can also be the result of deficiencies of other B vitamins, especially niacin.

Riboflavin therapy

Riboflavin is most frequently employed in the treatment of hacks in the corners of the mouth and other minor skin problems. The usual recommendation is an increased consumption of milk, offal, eggs, vegetables, and yellow fruits. If this is not possible, riboflavin can be taken in tablet form, possibly in a combination of B vitamins.

The orthodox use of riboflavin is in the correction of a deficiency condition. The risk of overdosing with riboflavin is minimal. The upper limit is 3 g per kg bodyweight per day, which is several thousand times the RDA. Megadoses can, however, lead to numbness and itching.

13.   NIACIN

It is strange, but true — for the truth is always strange.

Lord Byron (1788–1824)

Niacin is water-soluble, and, like the other B vitamins, cannot tolerate being heated

up. Niacin can be formed in the body by the amino acid, tryptophan (see 2.10), but most of our daily requirements are covered by the intake from our diet. The best sources of niacin are grains, vegetables, liver, meat and milk.

This vitamin plays an important role in the carbohydrate, protein and fat metabolisms. In addition, niacin produces energy in the cells, it has a certain antioxidant effect and it helps to reduce high blood levels of cholesterol.

Niacin occurs in two different forms: as nicotinate and as nicotinamide (niacinamide). In this section I have chosen to use the term niacin, except where the term niacinamide is expressly required.

Niacin participates in a whole range of processes which are of great importance to food metabolism. Its most important role in the protection of the cells is related to its function in the antioxidant enzymes, NAD and NADP. They are responsible for the breaking down of glucose, the metabolism of fatty acids and the production of energy. Niacin is neither nicotinate nor nicotinamide, but both together. The amino acid, tryptophan, is partially transformed into niacin in the organism.

Niacin requirements and metabolism

The RDA for niacin is described either in Niacin Equivalents (NE) or in milligrams. The latter is the most commonly used description and the RDA is set at 19 mg for this vitamin. This corresponds to 19 NE or to 1.14 mg tryptophan. As is the case for the other B vitamins, the daily requirement for men is greater than that of women, because men usually have a greater calorie intake than women. Pregnancy and breast feeding nevertheless increase niacin requirements by 2 and 5 mg respectively.

Infants who are less than one year old have a daily intake of 6–8 mg, but, after the first year, requirements are more or less the same as for adults.

Niacin requirements are increased in the event of deficiencies of vitamins B1, B2 and B6, because these vitamins participate in the transformation of tryptophan to niacin. The transformation of tryptophan to niacinamide requires the presence of vitamin B6. It would appear that contraceptive pills may also inhibit the transformation of tryptophan into niacin.

One-third of niacin occurs as niacin, the other two-thirds arise as tryptophan. Apart from liver, meat, fish, milk, eggs and yeast, other good sources of niacin are avocados, figs and prunes. In practice, most of our niacin intake comes from liver and meat, while the bio-availability of the niacin in grains varies considerably. Maize contains small quantities of niacin and tryptophan and a great deal of this is in a form which human beings are unable to metabolize. In many industrialized countries niacin, thiamin and riboflavin are added to breakfast cereals.

The effects of niacin

Niacin improves the functioning of the digestive system, it protects the skin and it helps to improve bad circulation by expanding the arteries and thereby reduces resistance to blood flow.

The classical symptom of niacin deficiency is *pellagra*, a disease which has been more or less eradicated in the industrialized world. Nevertheless, it is still widespread in slum areas, amongst alcoholics and in underdeveloped countries. Characteristics of the disease are skin changes, infected mucous membranes and mental defectiveness. If the disease is not treated with niacin supplementation, it will eventually lead to death.

Niacin protects the cells from a number of harmful effects. Experimental animals have been given the weed-killer paraquat, which increases the quantities of superoxides. The harmful effects of this substance are counteracted by niacin, which may have other antioxidant properties in the organism.

High levels of dangerous LDL-cholesterol, the oxidized o-LDL form of which increases the risk of hardening of the coronary arteries, can be reduced with niacin, suggesting that niacin may be a contributory factor in the prevention of heart disease.

In many ways niacin (nicotinate) can be considered as the counterpart of nicotine. Although the chemical structure of these two substances resemble each other, they have diametrically opposed effects in the organism. Nicotine causes the blood vessels to contract while nicotinate causes them to expand. Nicotine increases the formation of lipids (fats) in the blood and nicotinate reduces lipids and actually dissolves accumulations of fats.

Niacin therapy
During pregnancy and breast feeding, niacin intake should be somewhat larger than normal. Niacin is used in the treatment of contracted blood vessels and dizziness, and can also be employed, together with other vitamins in the B group, to reduce the loss of fluids associated with burns.

Niacin is given (in 10-fold doses) to patients suffering from disorders of the peripheral nervous system (disturbances in the sense of touch). The tranquilizing effect of niacin is exploited in the treatment of migraine, alcoholism and schizophrenia in daily doses of 3000 mg, combined with 3000 mg of vitamin C. In the USA in particular, niacin is used as a substitute for sleeping pills, although this effect has not been acknowledged by many doctors. Tryptophan, the precursor of niacin, would, on the other hand, appear to be well-suited to this purpose (see 2.10).

Nicotinate can have certain side-effects. Very large doses can give a flushed feeling in the skin, which reddens, and it can produce feelings of dizziness and general discomfort. These symptoms are caused by the expansion of the blood vessels. Side-effects can be prevented by starting with a low dosage and gradually building up to larger doses over a period of time.

Niacinamide does not produce the same strong side-effects, but, on the other hand, this substance does not produce the same cholesterol-reducing effect.

Nicotinate can produce a short-lived feeling of fatigue, and, because it is an acid, it can give rise to minor stomach complaints. In order to avoid these effects on the stomach one can take nicotinate together with meals or with acid neutralizing medicines, which do not disturb the effects of niacin.

A daily intake of more than 3–6 g over an extended period can give rise to skin changes. Niacinamide has a tendency to upset the sugar balance, causing the sugar levels in the blood to rise. This increase in blood sugar only lasts until intake is reduced.

14. PANTOTHENATE

Science is a splendid piece of furniture to have on the first floor, as long as there is good sense on the ground floor.

O. W. Holmes (1809–1894)

One can find pantothenate in virtually all living creatures. The best sources of this vitamin are egg white, liver and yeast. Pantothenate is a component of coenzyme A, which plays a vital part in the carbohydrate, protein and fatty acid metabolisms. Pantothenate deficiency leads to circulatory disturbances, rhinitis and disturbances in the digestive and nervous systems. This vitamin is also an antioxidant and is beneficial to the functioning of the B-lymphocytes — the cells which are responsible for the production of antibodies. Pantothenate is therefore a vital building brick in the cell defence against almost all diseases.

All animals, plants, bacteria and yeasts need pantothenate in order to survive. This is why this essential vitamin is found in almost all foodstuffs. Its importance is reflected in the Greek root of the name: *panthos* means everywhere.

Originally this substance was called vitamin B3, but from 1938 it has been called pantothenate, according to the wishes of **J. R. Williams**, who discovered this vitamin. Pantothenate participates in coenzyme A and through this it has a distinct influence on carbohydrate, protein and lipid metabolism. There are no specific diseases caused by dietary deficiencies of this vitamin.

Pantothenate requirements and metabolism

Adult humans usually have a daily intake of 5–20 mg of pantothenate. This quantity is considered ample, also for pregnant women, although precise requirements are not known. Some textbooks recommend 4–7 mg, while others suggest as much as 10–50 mg.

Intestinal bacteria apparently produce pantothenate, but otherwise the best dietary sources are offal, egg white, cheese, vegetables, bean sprouts, bran, peas, and brewers' yeast. Broccoli, lean meat, milk, potatoes and syrup also contain this vitamin.

In contrast to many of the other water-soluble vitamins, pantothenate is quite heat-resistant and is not destroyed by cooking, baking or frying. On the other hand, it cannot tolerate the high temperatures generated in a pressure cooker, and it is also partially broken down in deep-frozen meat. Pantothenate is partially absorbed in the stomach while the rest is assimilated in the small intestine. As a participant in coenzyme A, pantothenate is distributed throughout all the tissues in the organism, with the most substantial concentrations accumulating in the liver, kidneys and heart. Blood levels for this vitamin are usually about 100–400 μg l^{-1}. The amounts of pantothenate which are eliminated are usually equal to intake. Two-thirds are transported out of the body with the urine, while the other third is eliminated with the excreta. Pantothenate is apparently not broken down in the organism, nor is it stored in the organism to any significant degree. This is why we require a constant supply from our diet.

The effects of pantothenate

Apart from its role in the production of energy from the raw materials, sugar, protein and fat, pantothenate also contributes to the formation and growth of new cells and to the production of essential fatty acids. This vitamin is also required, in order for the body to be able to utilize para-aminobenzoic acid (PABA) and choline, and for the normal functioning of cortisone. In addition, pantothenate is of significance in the production of antibodies.

Pantothenate deficiency only occurs in human beings under experimental conditions, where volunteers are given a pantothenate-free diet, or where they are given antagonists which counteract the activity of pantothenate. In this situation, the volunteers can experience symptoms of fatigue, sleeplessness, personality changes, discomfort, stomach pains, numbness in the extremities, muscle cramps and reduced antibody production. All of these symptoms disappear when a supplement of pantothenate is given.

It has been claimed, from the results of experiments with rats, that pantothenate can prevent the hair from turning grey, or that it can even restore the original hair colour, but there is no evidence to show that this effect also applies to human beings.

Pantothenate therapy

Sodium and potassium salts of pantothenate are used medicinally in the treatment of circulatory disorders in the legs, or so-called restless legs. This form of treatment is based on the discovery that a pantothenate-free diet produces such symptoms. There have been cases where this treatment has been successful, but no controlled studies have been conducted into this phenomenon.

Calcium pantothenate has been used with some success in adjunct therapy for paralytic ileus, an intestinal disorder which frequently appears after operations, where the patient has been unable to move for a long time. Dosages for this treatment are very high, up to 500–1000 mg, and are administered in the form of an injection. Burns and other skin injuries are sometimes treated with pantothenate ointments. Pantothenate also appears to improve the activity and peristalsis of the small intestine.

Healthy people are unlikely to need pantothenate supplements, even though these are harmless.

15.   PYRIDOXINE (VITAMIN B6)

A truth usually has a life span of about 17–18 years, or 20 at the most, very rarely does it last longer than this.

Henrik Ibsen (1828–1906)

The incidence of subclinical vitamin B6 deficiency is high amongst the elderly (up to 25–30%). Vitamin B6 is required for general resistance to disease and premature ageing, for the production of new red blood cells and for the use of selenium. Pyridoxine is also essential to the maintenance of the skin and the nerves, and it helps to inhibit the aggregation of blood platelets (thus reducing the risk of blood clots). It

also improves the activity of the T- and B-lymphocytes in the immune system, thus contributing to the prevention of asthma, allergies, arthritic disorders, cancer, coronary heart disease, circulatory disorders and a range of other ailments. These 'new' effects of this vitamin have only been discovered very recently.

Pyridoxine requirements and metabolism
Requirements for pyridoxine are worked out in relationship to the metabolism of the amino acid, tryptophan. Requirements are set in the region of 0.02 mg per gram of protein. According to these calculations, women would need 2 mg pyridoxine per day and men 2.2 mg per day. Recommendations for infants up to one year old are 0.3–0.6 mg, and for older children and adolescents 1–2 mg per day. During pregnancy and breast feeding the daily requirement (RDA) is estimated at about 2.5 mg per day.

Pyridoxine requirements are presumably higher for those who take contraceptive pills. In 10–20% of women on the pill, biochemical indicators show pyridoxine deficiency. Another risk group involves those who are on a high-protein diet, including athletes. Requirements are also greater for people who suffer from chronic illnesses, and particularly for those who suffer from intestinal diseases, such as gluten allergy.

My own studies, conducted in a Finnish nursing home, showed that 25–30% of the residents suffered from pyridoxine deficiency. This deficiency condition was corrected by a daily supplement of 2 mg of pyridoxine.

The best natural sources of pyridoxine are meat, kidneys, liver, egg white, milk, yeast, grain husks and green vegetables. Vitamin B6 is destroyed by being heated up, by contact with light and by long-term preservation in tins or in a deep-freeze. The latter destroys about 20% of the pyridoxine in a given foodstuff. Frying destroys about half of the pyridoxine in food, while refined grains lose up to 90% of this important vitamin.

Pyridoxine is easily absorbed in the intestines and has a rapid metabolism. Only five hours after sufficient intake, the urine concentrations of this vitamin are optimal. The absorbed pyridoxine is found in cells throughout the body where it has a range of important functions.

Pyridoxine values in whole blood, red blood corpuscles and in plasma can be measured and these reflect the levels of the organism's reserves of this vitamin. Certain enzymes (s-ALAT, s-ASAT) measured in the blood also give a picture of vitamin B6 status for humans.

The effects of pyridoxine
Vitamin B6 is essential to the immune defence (production of antibodies). It also inhibits the production of harmful prostaglandins and increases concentrations of beneficial prostaglandins throughout the tissues. In addition, pyridoxine contributes to the prevention of the aggregation of blood platelets and thus contributes to the inhibition of blood clots. Vitamin B6 may also protect us from arteriosclerosis and heart attacks by limiting the accumulation of plaques in the artery walls. This has been observed in monkeys and rabbits, and it is now believed that it may also apply to human beings. Essential fatty acids also exhibit this effect (see Chapter 6).

In fact, vitamin B6 and essential fatty acids (EFAs) seem to have a lot in common.

The skin changes in vitamin B6 deficiency resemble those of EFA deficiency, and they can be cured by vitamin B6 supplementation and vice versa: EFA supplementation cures skin symptoms caused by B6 deficiency. The metabolic mechanisms for these events are not yet understood.

Vitamin B6 also interacts with cholesterol in the body: in most cases a daily supplement of 400 mg pyridoxine decreases high cholesterol levels in the blood in apes and in humans.

Vitamin B6 also participates in the metabolism of sugars and proteins and is also of importance for the normal functioning of the nervous system, the muscles and the heart. Deficiencies of this vitamin can lead to reduced resistance to disease and senile decay.

Vitamin B6 ensures the production of niacin and haemoglobin. This is why deficiencies can lead to microcellular anaemia. In this form of anaemia, the iron content of serum is high, but the organism is unable to utilize it. This condition can lead to leukaemia. Finally, vitamin B6 may be vital for the use of the trace element, selenium.

Low levels of vitamin B6 have been recorded in cases of acute heart attacks. Symptoms of vitamin B6 deficiency are loss of appetite, emaciation, nervousness, fatigue, depression, poor memory, sleeplessness and general apathy. Hacks in the corners of the mouth, infections of the oral cavity and loss of the sense of touch in the extremities can also be symptoms of vitamin B6 deficiency, although these symptoms often indicate that the patient is also deficient in other B vitamins, B2 in particular. In children, severe deficiencies of this vitamin often express themselves as irritability and epilepsy-like convulsions. Some people suffer from a hereditary condition which increases their pyridoxin needs. This may result in convulsions in the first few months after birth, but this can be arrested by pyridoxine supplementation.

People with a high alcohol consumption are particularly susceptible to pyridoxine deficiencies, as are women who use contraceptive pills. Pyridoxine deficiency may also associate with pregnancy and menstruation discomfort. Symptoms of this may resemble those of pre-menstrual tension, such as depression and bad moods. These symptoms can often be alleviated with supplements of vitamin B6, although this has not been confirmed in all controlled studies. In addition, an extra supplement of pyridoxine may be able to stabilize the blood sugar balance during pregnancy or while contraceptive pills are used.

Pyridoxine therapy

Many forms of medicine are antagonists to vitamin B6, i.e. they counteract its effects. This applies first and foremost to INH (tuberculosis medicine) and penicillamine (anti-arthritis medicine). These medicines can lead to serious deficiencies of vitamin B6, resulting in symptoms of drowsiness, loss of sensation in the extremities, difficulties in walking, and, in severe cases, to convulsions. These side-effects can be alleviated or prevented with 30–100 mg of pyridoxine per day. For symptoms in the nervous system, as much as 300–600 mg per day are sometimes given.

I recommend a supplement of 2 mg of vitamin B6 per day to my elderly patients. In the event of asthma or intestinal diseases, such as gluten allergy or cancer, or after major intestinal surgery, a daily supplement of about 50–150 mg can be appropriate.

I often prescribe 100–300 mg per day for patients with angina pectoris or as adjunct therapy after a heart attack. The effect is reinforced when pyridoxine is combined with selenium, as the risk of blood platelet accumulation (and the concomitant risk of another blood clot) is reduced. A French study has recently shown that asthma patients experienced an improvement in their condition after a daily supplementation with 100 mg vitamin B6.

Inflammation plays a crucial role in asthma, and these micronutrients, taken simultaneously, counteract inflammation. According to my experience, the effect is enforced when pyridoxine is given together with EFAs and antioxidant vitamins and minerals. This is due to the fact that sensitive patients may experience side-effects at a dosage of 600 mg pyridoxine per day. Symptoms can be stomach pains, changing moods, discomfort, dizziness, sleeplessness and sweating. These symptoms can usually be held in check by reducing or halving the dosage.

Warning: Daily doses of more than 1000 mg can be dangerous. Cases of poisoning have been reported in the USA and Australia after such megadoses, which have caused damage to the peripheral nervous system and have resulted in disturbances in the sense of touch.

16.   CYANOCOBALAMIN (VITAMIN B12)

The truth is all too naked. It does not titillate.

Jean Cocteau (1891–1963)

Vitamin B12 deficiency leads to the serious blood disease, pernicious anaemia. This deficiency disease is not caused by insufficient intake of vitamin B12 from diet, but by malabsorption. Pernicious anaemia is treated with vitamin injections at intervals of several months throughout life.

Vitamin B12 is required for the production of red blood corpuscles and for normal growth. At the same time, this vitamin improves resistance through its influence on the activity of the T- and B-lymphocytes in the immune system. Adjunct therapy with folate and vitamin B12 may be able to inhibit cancers and limit the proliferation of metastases.

Vitamin B12 requirements and metabolism

Daily requirements of this vitamin are very small — only 3 μg. During pregnancy an extra daily microgram is recommended. Requirements of this vitamin are also increased in connection with an overactive thyroid gland and pyretic diseases (fevers), i.e. situations in which the metabolism functions more rapidly.

Patients who suffer from pernicious anaemia and megaloblastic anaemia need more vitamin B12 than others, and this is usually administered in the form of injections, because the vitamin cannot be absorbed in the intestines.

The low levels of vitamin B12 requirements have been illustrated in a study on rats, conducted by Dr **Antti Ahlström**, professor of nutrition at the University of Helsinki. Two groups of rats were fed with a diet which was low in this vitamin, but only one of the groups developed deficiencies. It was later discovered that this was because one group has been fed by hand, while the other group was fed with a spoon. The group which had been fed by hand did not develop deficiencies because the fodder was able to absorb sufficient quantities of vitamin B12 from the sweat on the researcher's palm!

The best natural sources of vitamin B12 are liver and kidneys, although it is also found in milk, eggs, cheese, meat, fish, prawns and oysters. Plants absorb this vitamin from composted fertilizers and it is also produced by bacteria in the roots of peas. Many textbooks on medicine and nutrition have stated that vitamin B12 does not occur in plants, but this is incorrect.

A strict vegetarian diet, excluding eggs and milk, can sometimes mask the presence of vitamin B12 deficiency. The disease resembles that of folate deficiency, in that it involves changes in the blood and disturbances in the nervous system.

A normal mixed diet contains 7–30 μg of vitamin B12. Only 1–1.5 μg of this total intake is absorbed into the blood. In order to be absorbed, vitamin B12 requires the presence of another substance, known as the *intrinsic factor*. The vitamin, together with the intrinsic factor, passes through the intestinal mucous membranes and is then released into the bloodstream. The vitamin's passage through the intestinal wall also requires the presence of calcium, and if the intrinsic factor is not present in the stomach, the vitamin is not absorbed. In such a case, it is possible to administer large doses orally in the hope that at least some of it will be absorbed.

Several different diseases and medicines can affect the complex process whereby vitamin B12 is absorbed. Four to six per cent of patients who have undergone major stomach surgery have difficulties with B12 absorption for as long as 10 years after the operation. Various intestinal infections and inflammations can also affect B12 absorption, as can certain cytostatics (vinblastine) and cortisone. There are several chemotherapeutic medicines which destroy the intestinal mucous membranes themselves, and therefore prevent absorption. These include cyclophosphamide, methotrexate and 5-fluorouracil.

It is a relatively simple matter to check vitamin B12 content in plasma. Normal values are 200–900 picomoles per litre. The vitamin B12 which is absorbed is found in all the cells in the body, but 90% of it is stored in the liver.

Some of this vitamin is eliminated with the bile, the rest with the excreta. Elimination of vitamin B12 is a slow process and the half-life in the liver is about one year. Intestinal capacity for absorbing vitamin B12 can be checked using the so-called Shilling Test: healthy individuals absorb about 30% of the test dose, while those who suffer from pernicious anaemia only absorb about 2%.

Vitamin B12 therapy
Pernicious anaemia is a very rare disease for children, but more common for the elderly, especially in Northern Europe, where an hereditary pattern has been detected for the disease. If the stomach is surgically removed, the same conditions arise as for patients who lack the intrinsic factor: there are no longer any mucous membranes which can secrete the factor. Another operation which can disturb vitamin B12 absorption is removal of the end of the small intestine (e.g. in Crohn's disease), where absorption in the bloodstream takes place.

Vitamin B12 deficiency can also arise in connection with long-term diarrhoea, gluten allergy or gastroenteritis. These conditions are usually also associated with folate deficiency.

Vitamin B12 deficiencies can appear either as large-cell anaemia, tongue infections (the tongue often looks shiny and bright red) or disturbances in the nervous system (loss of sensation and poor coordination). If the condition continues, the patient's general health becomes weakened and this can lead to symptoms in the heart and the circulation, the skin becomes yellowish in colour and the patient suffers from diarrhoea and other stomach problems.

Patients with pernicious anaemia often have additional mental symptoms of depression, irritation, loss of concentration and paranoia. These symptoms occur in 25–30% of patients. Over 60% of patients also exhibit changes in their EEG patterns (electroencephalogram — a measure of electrical impulses in the brain). This may be caused by damage to the myelin in the nerve endings. Conversely, it has been discovered that 2% of all psychiatric patients suffer from a previously undiagnosed pernicious anaemia.

In the past, the treatment for pernicious anaemia involved the patient in eating large quantities of liver every day. The current treatment for pernicious anaemia usually consists of vitamin B12 injections, in large doses of up to 1000 μg, which covers requirements for about three months at a time. Pernicious anaemia can also be treated with very large doses in tablet form, but these are more expensive and less reliable. Injections, on the other hand, invariably give good results.

17.   FOLIC ACID (FOLACIN, FOLATE)

The great truths are always simple. Complicated truths often contain a grain of a lie.

Maria Jotuni (1880–1943)

Folate is found in all green plants, in liver and in yeast. Normally, human intestinal bacteria are able to produce sufficient folate. Folate and its derivates take part in the formation of nucleic acids and are therefore vital to the formation of new cells, the growth of the embryo and normal development. Folate deficiency can lead to large-cell anaemia.

Various medicines, which are used in the treatment of problems in the urinary system and cancer, are antagonists to folate. Users of contraceptive pills, epilepsy medicines and alcohol are particularly susceptible to folate deficiency.

Folate deficiency often accompanies vitamin B12 deficiency. Folic acid was discovered by Dr **Lucy Wills** in Bombay, India, in 1931. She was doing research into large-cell anaemia, which was a common complaint amongst pregnant women. She conducted an experiment on monkeys, in which she fed them on a diet of white rice and white bread, the same diet as her patients received, and the monkeys developed the same symptoms as her patients. Dr Wills tried all the known vitamins of the time, as well as liver extract, without success. She discovered, however, that yeast could cure the disease and deduced that yeast must contain a previously unknown substance which could cure and prevent large-cell anaemia.

This 'Wills Factor' was isolated from spinach leaves in 1941, and was given the name folic acid (folate). The name is derived from the Latin word *folium*, which means leaf. Folate came into use in 1945 for the treatment of large-cell anaemia and later it was used in connection with various forms of tropical anaemic diarrhoea.

It is estimated today, that folate deficiency is the commonest vitamin deficiency condition in both the underdeveloped and industrialized parts of the world. It has been claimed that 45% of the lower social groupings in the United States suffer from folate deficiencies. In the western world, however, this deficiency is commonest amongst the elderly. Yet folic acid is rarely used therapeutically even in the best current medical practice.

Folate requirements and metabolism
About 25–50% of the dietary folic acid is biologically available. The minimum requirement for human beings is about 50 μg and the RDA is set at 400 μg. During pregnancy, a doubling of this level of intake is recommended, i.e. 800 μg per day, and during lactation 500 μg is recommended. Requirements are doubled during pregnancy because the foetus needs folate for the formation of new cells. The RDA for children is 10 μg per kg bodyweight.

A well-balanced diet provides 250–700 μg folate per day, according to figures from different parts of the world. In Britain, one study showed an average intake of 670 μg per day. The best natural sources of folate are vegetables, asparagus, peas, whole grains, nuts, liver, kidneys and yeast. Breast and cow's milk contain 2–3 μg per 100 ml, which usually meets the needs of the baby. Folate (e.g. in milk) is unable to tolerate being cooked and this is why infants on pasteurized or sterile milk need folate substitution. Human beings have very small reserves of folate, and insufficient intake can lead to deficiency symptoms in the space of 3–4 months. The elderly on unvaried or inadequate diets, those who fast, pregnant women, those who take contraceptive pills and those who have a high consumption of alcohol are particularly exposed to the risk of folate deficiency. A combination of alcohol and contraceptive pills, or alcohol and pregnancy, further increases this risk.

Folate in the diet is absorbed rapidly and completely in the small intestine, even at dosages of up to 15 mg. At higher doses than this, the excess will not be absorbed. Of the folate which is absorbed, a proportion is eliminated with the bile into the intestines, from where a considerable proportion is then reabsorbed into the bloodstream. This circulation spares the folate reserves in the organism and helps to regulate the folate balance, which is highly sensitive to disturbances in the production of bile. Folate is eliminated with the excreta, and only when intravenously injected is it eliminated in the urine.

It is possible to investigate the folate balance in the organism by measuring the concentrations of folate in whole blood, red blood corpuscles and serum. One can also use a special histidine test in which the levels of formiminoglutamate (FIGLU) are measured. Folate deficiency is accompanied by an increase in the levels of this substance.

Folate deficiency
Folate deficiency anaemia (megaloblastic anaemia) can occur in children under two years of age who are not breast-fed, during pregnancy, after long-term diarrhoea or because of other disturbances in absorption of this vitamin (malabsorption). This form of anaemia occurs in all of 20% of expectant mothers. In developing countries a very widespread form of anaemia occurs, which is also caused by folate deficiency. In

addition, concealed folate deficiency is believed to be very common amongst pregnant women in developing countries.

If the foetus is exposed to folate deficiency, this can give rise to deformities and damage to the central nervous system and the brain. A deformity such as a hair lip or cleft palate may be connected with folate deficiency.

High alcohol consumption, long-term loss of appetite and a range of chronic diseases, including gluten allergy, may also lead to folate deficiency.

A British study has shown that 24% of surgical hospital patients suffer from folate deficiency, while other studies indicate that as much as 70–80% of the elderly in institutions are also deficient in this vitamin. They exhibit symptoms of apathy and depression. Other symptoms of folate deficiency are infections in the oral cavity and on the tongue, dizziness, respiratory problems and a grey discoloration of the skin. Similar skin changes have also been observed in pregnant women and users of contraceptive pills.

Epilepsy patients also run a considerable risk of developing folate deficiencies because the epilepsy medicines, phenytoin and phenobarbital, reduce the absorption of this vitamin.

Folate supplementation should perhaps be made routine for those who use contraceptive pills and epilepsy medicines.

Folate therapy can sometimes help in the correction of megaloblastic anaemia, which stems from diseases of the small intestine, e.g. diverticulitis.

Folate therapy

Previously, children who suffered from folate deficiency anaemia were given 5 mg of folate three times a day. Recent research has indicated, however, that the treatment can be effective at much smaller doses of about 25–100 μg per day. The patient's condition improves within a few days and appetite returns, while the blood balance is restored over a relatively short period. Patients who have had the stomach surgically removed should be tested for the levels of folate in the red blood corpuscles, and serum values of vitamin B12 should also be checked over a period of five years after the operation.

Embryo damage can perhaps be avoided by folate supplementation during pregnancy, although it is best if the supplementation is started before pregnancy, because any damage which occurs usually takes place in the early months.

Multivitamin products usually contain 0.1 mg (=100 μg) of folate. This maximum dosage is set for safety, because large doses of folate can conceal pernicious anaemia. I normally prescribe a 5 mg daily supplement of folate for my rheumatoid arthritis patients, in order to raise their production of red blood corpuscles.

18. BIOTIN

What he said was not true, but nevertheless it was quite right.
                                                Gustav Lindborg (1875–1929)

Biotin participates in the carbohydrate, protein and fat metabolisms. This vitamin occurs naturally in yeast, egg yolk and milk, but the intestinal bacteria produce so

much biotin that this covers our requirements. Nevertheless, a special course of antibiotics for children can inhibit the absorption of biotin in the intestines.

Some years ago a fasting diet based mainly on raw eggs was very popular. It turned out that this diet could lead to biotin deficiency, because raw eggs contain a glycoprotein (avidin) which limits the absorption of biotin. Egg consumption needs to be in excess of 15–30% of total energy intake before this condition occurs, however (this corresponds to about 20 eggs per day). This deficiency can be easily corrected with supplements of biotin. When an egg is cooked, the avidin is transformed in such a way that it is no longer able to form compounds with biotin.

Biotin, through its contributory role in the production of antibodies, helps to increase the effectiveness of the immune system. There are two known deficiency diseases associated with biotin.

Biotin was discovered in the same way as the other B vitamins and was first isolated in 1936. The word biotin is derived from *bios*, a factor which helps yeast to rise.

Biotin requirements and metabolism

Average biotin intake is 100–300 μg per day. A normal, mixed diet contains plenty of biotin, which is also produced by intestinal bacteria. Total quantities of excreted biotin are usually 3–6 times the amount of biotin intake from food.

Under normal circumstances, biotin deficiency is very rare. Biotin occurs naturally in liver, kidneys, meat, egg yolks, milk, cauliflower, nuts, chocolate and yeast. In addition, most vegetables contain small quantities of biotin.

The effects of biotin

Biotin is one of the most active existing compounds. As little as five-thousandths of a microgram stimulates the growth of yeast and bacteria. Biotin in food and in our tissues is a component of a protein-containing enzyme system (biocytin). Biotin operates as a coenzyme in a number of metabolic processes.

An experimentally induced biotin deficiency in human beings (caused by a diet of raw eggs) gives the following symptoms: rashes, infections in the mucous membranes in the mouth, loss of appetite, discomfort, sleeplessness, depression, muscle pains and increased blood levels of cholesterol. Biotin supplements, in the form of injections of 150–300 μg per day, are able to clear up these symptoms in a matter of a few days.

The medical literature provides a number of descriptions of biotin deficiency (in connection with diets of raw eggs). The elderly, athletes and epileptics have been shown to have relatively low blood values of biotin.

Although no 'natural' deficiency diseases associated with this vitamin have been discovered in adults, there are two known skin diseases which affect children, which have been linked with biotin deficiency. One is called Leiner's disease, which is a widespread form of seborrhoic eczema. The other disease is a fungoid infection which affects the growth of hair and results in disorders of the central nervous system and the immune system. In this case, the problem is associated with a failure in the function of the carboxylase enzyme, which is dependent upon supplies of biotin. When children suffering from these diseases are given a daily dose of 5 mg of biotin, the symptoms disappear quickly.

Both of these aforementioned diseases are extremely rare. Normal seborrhoic eczema has no connection with biotin deficiency. Attempts to treat a range of different skin diseases with biotin have therefore been unsuccessful. For patients with hair loss, it may be worthwhile to try a daily dose of 200–400 μg of biotin, together with para-aminobenzoic acid (PABA), inositol (see below), minerals and essential fatty acids.

19. VITAMIN-LIKE SUBSTANCES

A number of substances have vitamin-like functions in the organism, but do not satisfy the criteria for an actual vitamin. They are presumably not entirely essential for the organism, which is usually able to produce them itself. Alternatively, these substances often occur in such large quantities in our diet that they are not considered so vital for survival. The most important vitamin-like substances are choline, inositol, coenzyme Q (ubiquinone), lipoic acid, para-aminobenzoic acid (PABA) and the bioflavonoids.

Choline
Choline is a much-discussed, vitamin-like substance. It is a component of lecithin, which participates in lipid metabolism. This activity of choline is shared with inositol. Choline passes unaffected from the blood to brain tissues, and there are studies which indicate that it improves memory. It is also claimed that choline can reduce high cholesterol levels in the blood, but this effect has not yet been conclusively proved. Brains, offal, wheat bran, brewers' yeast and egg yolk are good sources of choline. Choline is also to be found in lecithin, vegetables and milk (although there is virtually none in fruit).

Choline was first discovered in the 1930s and is not an actual vitamin because the organism can produce this substance from the amino acid, methionine. The content of choline in the body is high, compared with the small quantities of vitamins.

Choline requirements and metabolism
Choline is an essential substance, but minimum requirements are not known. Choline deficiency scarcely occurs in human beings, but the production of choline demands a number of preconditions: there must be an excess of methionine, vitamin B12 and folate. It only requires that one of these factors is missing, for the choline levels in the tissues to fall.

The effects of choline
Choline deficiency in rats leads to fatty liver and cirrhosis of the liver. Choline appears to prevent these conditions from developing in rats, and it also prevents fatty liver in dogs. In other experiments with animals, choline deficiencies have been associated with kidney damage, increased cholesterol and higher blood pressure.

These experimental results have not been repeated for human beings: there is no evidence to show that choline prevents liver disease, while the cholesterol-reducing properties of this substance are also disputed.

Nevertheless, choline deficiency has been suspected as a possible factor in arteriosclerosis, and, possibly, in Alzheimer's disease. The latter involves a rapid ageing of brain tissue, causing the patient to become prematurely senile. Choline supplements have not, however, had any observable effect on this disease, for which no cure is known.

Choline Therapy
At the present moment in time there is no scientific basis for the medicinal application of choline. Some doctors may recommend choline on the basis of their own personal experiences (improvements in memory and speech impediments). Choline has been the object of fairly rigorous research in the field of neurological diseases but the results have not been very convincing.

In some people, preparations of choline can cause changes in the activity of intestinal bacteria, giving the excreta a strong smell of rotten fish. Choline bitartrate can cause diarrhoea, while choline chloride (choline hydrochloride) does not produce this side-effect.

Inositol
Inositol is another of the vitamin-like substances which play a role in lipid (fat) metabolism. It occurs naturally in liver, brewers' yeast, wheat bran, whole grains, maize, nuts, fruit, milk (especially breast milk), oats and syrup. It is also found in smaller quantities in cabbage, raisins, peanuts and grapefruit.

Like choline, inositol is a structural component of lecithin. Viewed chemically, inositol is an alcohol, but because of the molecular ring structure inositol is also related to sugars. This substance was isolated already in the 1800s from the urine of diabetics.

Inositol occurs in no fewer than nine different forms, but myoinositol is the only form which is biologically active and which is of importance in the metabolism. Inositol is not an essential substance and its exact function in the body is not yet known.

Inositol requirements and metabolism
In nature, inositol occurs in at least four forms: free inositol, phytate (inositolhexa-phosphate salts of magnesium and calcium), phospholipid- and phosphatidyl-inositol and in the form of a water-soluble compound. Large quantities of phytate in grains and unleavened bread bind calcium and can inhibit the absorption of various minerals, especially zinc. Phytate can, however, be broken down by the enzyme, phytase.

Average daily requirements are about one gram, although no actual RDA exists.

Inositol is absorbed from the intestines. A proportion of it becomes sugar, and some of it is led to the heart muscle, other muscles, the brain and other organs. Inositol levels in plasma are usually about 5 mg l^{-1}. There is very little inositol in the urine of healthy people, while inositol levels are increased in the urine of diabetics.

The effects of inositol
Experiments involving mice have shown that inositol deficiency in mice leads to loss of hair, reduced growth and milk secretion, rashes, constipation and congenital eye defects. For mice, inositol is a B vitamin, because it is an essential dietary factor for them.

Animal experiments have shown, moreover, that inositol has a choline-like effect on the metabolism — i.e. inositol inhibits the accumulation of fat in the liver and other organs and it participates in the construction of cell membranes and plasma lipoproteins.

Human deficiencies of inositol are not known and the substance has no significance for the treatment of metabolic disorders. In this instance, the results of animal experiments cannot be used to draw comparisons with human beings.

Ubiquinone (coenzyme Q10)

This biologically active coenzyme occurs everywhere in nature, hence the name (the prefix *ubi* means 'everywhere' in Latin). The term ubiquinone is used to describe a large group of different substances. Ubiquinone is found in the heart muscle, in yeast and in torula yeast.

The role of ubiquinone in the metabolism is now understood. It participates in a variety of *energy-producing* processes in the cells. Ubiquinone molecules are to be found in the innermost of the two membranes around the mitochondria, the 'power station' in the cell. The mitochondria contain a number of enzymes, known as the respiratory chain, in these membranes, which transport electrons from the metabolism to oxygen atoms during the production of energy and ubiquinone is involved in this process.

Ubiquinone is also an important *antioxidant* in the cells, which seems to be effective in protecting heart patients from reperfusion injury. In Japan, many heart patients take daily supplements of ubiquinone, prescribed by their doctors. In Japan, ubiquinone is registered as a medicine. I recommend 30 mg per day to my patients.

Para-aminobenzoic acid (PABA)

PABA is a component of the folic acid molecule and is therefore essential to microorganisms. Originally PABA was considered to be one of the B vitamins, because it participates in the synthesis of folate in the organism. In rats and mice, PABA is able to correct folate deficiencies totally. Human beings produce PABA from intestinal bacteria.

Chemically, PABA resembles sulpha preparations. This explains why PABA counteracts sulpha preparations and renders them ineffective.

PABA can prevent the hair of experimental animals from turning grey and can even restore normal hair colour, if this has been caused by vitamin deficiencies (e.g. biotin, folate or pantothenate deficiency). Unfortunately, PABA is unable to work this miracle in human beings. PABA is non-toxic, except in very large doses which can lead to discomfort and vomiting. There is, however, no medical reason for using PABA, except for its sun-protective properties, which are utilized in sun-tan lotions.

Bioflavonoids (vitamin P, rutin, hesperidin)

Bioflavonoids are a group of flavonoids which occur naturally in plants. In experimental animals, these substances strengthen the walls of the smallest blood vessels. Bioflavonoids are found everywhere in orange and yellow pigment, e.g. in rosehips, citrus fruits, vegetables, plums and bark.

Human requirements are not known and there is therefore no RDA. Human beings are scarcely in need of daily supplements of these substances and a possible requirement is fulfilled by small quantities of raw fruit and vegetables.

Some studies have, however suggested that it may be wise to take 100 mg bioflavonoids per 500 mg vitamin C, because the former improve the absorption of vitamin C in the intestines and reinforce the beneficial effects of vitamin C on the connective tissues.

Athletes are often given bioflavonoids because these are believed to speed up the healing of pulled muscles, sprained joints and chafed skin.

Rutin functions as an *antioxidant*. In the Soviet Union, rutin is viewed as a very promising antioxidant vitamin. An effective daily dose is about one gram.

20.   PANGAMATE AND AMYGDALIN

A half truth is a lie. If a truth is mixed with a lie, it becomes a complete lie.
Wilhelm Malmivaara (1854–1922)

Pangamate ('vitamin B15')

Pangamate is one of two much-disputed B vitamins, which were isolated from apricot stones for the first time in 1951. Since then, pangamate has been detected in rice husks, maize, other plants, and in brewers' yeast, ox blood and horse liver.

There are widely divergent opinions on the importance of this substance for our health. The discoverer of pangamate, **E. T. Krebs Sr**, his son, **E. T. Krebs Jr**, and one of their relatives, **Peter W. Stacepoole**, are quite convinced that the substance has almost all-powerful properties.

Chemically, this substance is glucono-dimethyl acetylate. It is claimed that this substance is an antidote to toxins, that it can cure cancer, alcoholism, hepatitis, diabetes and a whole range of other diseases.

Research into pangamate has been most active in the Soviet Union, while the American Food and Drug Administration (FDA) does not consider that there is enough concrete evidence for the beneficial effects of this compound. The FDA does not even classify pangamate as a provitamin, let alone a vitamin. Most doctors in the western world treat panganate as something of a confidence trick.

Amygdalin (laetrile, 'vitamin B17')

Amygdalin is taken more seriously by the scientific establishment, but it is neverthe-less a disputed substance. Amygdalin is the chemical term for laetrile, which occurs naturally in apricots and related fruits. It has been maintained that this substance has a cancer-preventive effect, and that it can actually be used in the treatment of cancer, a prerequisite being an antineoplastic (cancer-preventive) diet. This diet excludes meat and milk products completely, and consists mostly of fruits, nuts and extremely high doses of minerals.

The effect of laetrile is claimed to stem from the release of hydrogen cyanide in the tissues. However, hydrogen cyanide does not only destroy cancer cells, it also destroys healthy cells and is in fact one of the most potent known poisons. Some researchers are even of the opinion that laetrile can lead to cyanide poisoning.

Hundreds of studies have been conducted on laetrile, but these have produced conflicting results. One of the most comprehensive controlled studies was carried out in 1979 and failed to show any effect of laetrile on the progression of cancers. On the contrary, some researchers now even suspect that laetrile itself may be a carcinogen.

21. LECITHIN

Lecithin has become a popular food supplement in recent years, because there have been a number of indications that it may have a beneficial influence in the organism. Lecithin occurs in soya beans and vegetable oils, amongst other things, and is one of the phospholipids, substances which are also produced in the human liver. It is found in large quantities in human brain tissue (40% of the brain is composed of lecithin in the form of phosphatidyl choline).

When lecithin passes through the digestive system, it is decomposed into glycerol (an alcohol), phosphate, choline and fatty acids. The phospholipids are the building bricks in the mitochondria (the miniature power station in the cell) and membranes of living cells.

Phospholipids are required in the food metabolism for the transportation of lipids and for the production of energy. Two studies, one in Finland (by my research team) and one in Denmark, have indicated that lecithin may be able to decrease cholesterol levels and thereby the risk of arteriosclerosis, blood clots and ischaemic heart disease. A West German study, involving about 300 participants, has provided similar results. A number of non-positive studies exist, too.

Moreover, other studies have suggested that lecithin may improve memory and the mental well-being of the elderly, as well as having a potentially beneficial effect on certain neurological diseases, such as Alzheimer's disease, Friedrich's ataxia and myasthenia gravis.

LITERATURE

Cameron, E. and Pauling, L. *Cancer and vitamin C*. Warner books, 1981.

Chow, C. K., Thacker, R. R., Changchit, C., Bridges, R. B., Rehm, R. S., Humble, J. and Turbeck, J. Lower levels of vitamin C and carotenes in plasma of cigarette smokers. *J. Am. Coll. Nutr.* **5** (1986) 305–312.

Elevated dosages of vitamins. Benefits and hazards. International Symposium, Interlaken, Switzerland, 7–9 September 1987. Abstracts.

Friedrich, W. *Vitamins*. Walter de Gruyter. Berlin–New York, 1988, p. 1058.

Gerbershagen, H. U. and Zimmermann, M. (eds). *B vitamins in pain*. pmi Verlag, GmbH, Frankfurt, West Germany, 1988.

Gey, F., Brubacher, G. B. and Stähelin, H. B. Plasma levels of antioxidant vitamins in relation to ischemic heart disease and cancer. *Am. J. Clin. Nutr.* **45** (1987) 1368–1377.

Henneckens, C. H., Kimberley, Eberlein, M. P. H. A randomized trial of aspirin and β-carotene among US physicians. *Prev. Med.* **14** (1985) 165–168.

Himberg, J.-J., Tackmann, W., Bonke, D. and Karppanen, H. *B vitamins in medicine*. Helsinki 1985. Friedr. Vieweg & Sohn, Braunschweig/Wiesbaden, West Germany, 1986.

Korpela, T. and Christen, P. (eds). *Biochemistry of vitamin B6*. Birkhäuser, Basel-Toronto, 1987.

Lee, W. H. *Coenzyme Q*-10. Keats Publishing Inc., New Canaan, Connecticut, 1988.

Lemoyne, M., van Gossum, A., Kurian, R., *et al*. Breath pentane analysis as an index of lipid peroxidation: a functional test of vitamin E status. *Am. J. Clin. Nutr.* **46** (1987) 267–272.

Mathews-Roth, M. M. Photoprotection by carotenoids. *Fed. Proc.* **46** (1987) 1890–1892.

Mathews-Roth, M. M. Lack of genotoxicity with β-carotene. *Toxicol. Lett.* **41** (1988) 185–191.

Menkes, M. S., Combstock, G. C., Vuilleumier, J. P., Helsing, K. J., Rider, A. A. and Brookmeyer, R. Serum β-carotene, vitamins A and E, selenium, and the risk of lung cancer. *N. Eng. J. Med.* **315** No. 20 (1986) 1250–1254.

Ovesen, L., Ebbesen, K. and Olesen, E. S. The effect of oral soybean phospholipid on serum total cholesterol, plasma triglyceride, and serum high-density lipoprotein cholesterol concentrations in hyperlipidemia. *Parenteral Enteral Nutr.* **9** No. 6 (1985) 716–719.

Reynolds, R. and Leklem, J. (eds). *Clinical and physiological applications of vitamin B6*. Alan R. Liss, Inc., New York, 1988.

Simon-Schnass, I. and Pabst, H. Influence of vitamin E on physical performance. *Internat. J. Vit. Nutr. Res.* **58** (1988) 49–54.

Temple, N. J. and Basu, T. K. Does β-carotene prevent cancer? Critical appraisal. *Nutr. Res.* **8** (1988) 685–701.

Tolonen, M., Sarna, S., Knuutinen, V. and Seppänen, T. Lecithin can lower high blood cholesterol levels in men. A double-blind cross-over study using randomly chosen experimental subjects. *Holon* **7** (1988) 8–12.

Virtamo, J. J., Haapakoski, J., Huttunen, J. K., *et al*. Prevention of lung cancer with β-carotene and α−tocopherol. *Finlands Läkartidning* **42** No. 6 (1987) 456–462.

Wald, N.J., Thompson, S. G., Densem, J. W., *et al*. Serum β-carotene and subsequent risk of cancer: results from the BUPA study. *Brit. J. Cancer* **57** (1988) 428–433.

Wartanowicz, M., Panczenko-Kresowska, B., Ziemlanski, S., *et al*. The effect of α-tocopherol and ascorbic acid on the serum peroxide level in elderly people. *Ann. Nutr. Metab.* **28** (1984) 186–191.

5

Minerals and trace elements

1. WHAT IS THE SIGNIFICANCE OF MINERALS AND TRACE ELEMENTS FOR HEALTH?

Human beings are constructed in such a fashion that they ought to know everything.

Pentti Haanpää (1905–1955)

The importance of vitamins and minerals for our health has only become clear in very recent years. Selenium is a prime example of this. Nutritional experts and orthodox medicine have long ignored the importance of selenium, and doctors who prescribed selenium have more or less been treated as cranks. Attitudes are changing rapidly, however, and today most sceptical researchers and authorities are prepared to concede that selenium and other trace elements are of far greater importance to the organism than their insignificant quantities might lead us to believe.

At this very moment the interest in selenium research is growing rapidly and the risk groups for selenium deficiency are constantly being added to. Pregnant women and breast-feeding mothers, children and adolescents, the elderly, the sick undergoing constant medical treatment and people on unvaried diets are only some of the groups of people who are in the danger area.

All in all, there are 104 different elements. Human beings require at least 50 of these elements in their diet — they are vital to our well-being, just as vitamins are. These essential elements are found in the blood, in tissue, in organs and in bodily fluids. Other elements are alien to the organism and may represent a danger to health. Many medicines are composed of elements which are foreign to the organism.

No new vitamins have been discovered since the 1950s, but the list of essential trace elements continues to grow. The most recently discovered essential trace elements are selenium, chromium, fluorine, molybdenum and manganese. Minerals and vitamins represent only a very small proportion of our total body weight, only 4%, in fact. The other 96% consists of oxygen, hydrogen, carbon and nitrogen, with oxygen alone making up more than 50% of the total. Oxygen and hydrogen constitute more than two-thirds of our total body weight, first and foremost because water, which occurs in large quantities in blood and other bodily fluids, is composed from these two elements.

We often use the terms mineral element, trace element and mineral indiscriminately, when we describe the metallic elements which are components of our diets and of our bodies.

In a medical context, minerals are divided into three main groups, (according to the level of daily requirements): macrominerals, microminerals and trace elements.

Macrominerals
In this group are included the metallic elements of which our daily requirement is more than 100 mg. These are:

- calcium (chalk) (Ca)
- phosphorus (P)
- sodium (Na)
- potassium (K)
- chlorine (Cl)
- magnesium (Mg)
- sulphur (S)

Microminerals
The daily requirements of minerals in this group range from less than 1 mg and up to 100 mg. This group includes:

- iron (Fe)
- copper (Cu)
- zinc (Zn)
- manganese (Mn)
- iodine (I)
- molybdenum (Mb)
- selenium (Se)
- fluorine (F)
- bromine (Br)
- chromium (Cr)
- cobalt (Co)
- silicon (Si)

Trace elements
Actual daily requirements for these elements have not yet been set, but it is a question of micrograms.

- arsenic (very small quantity) (As)
- boron (B)
- tin (Sn)
- nickel (Ni)
- germanium (Ge)

- vanadium (V)
- tungsten (Wo)
- lead (very little) (Pb)

Alien substances
These substances are foreign to the organism, and their presence in the body can cause damage:

- lead (in larger quantities) (Pb)
- arsenic (in larger quantities) (As)
- cadmium (Cd)
- mercury (Hg)
- barium (Ba)
- aluminium (Al)
- radioactive elements
- lithium (Li)
- beryllium (Be)
- silver (Ag)
- gold (Au)
- antimony (Sb)
- and many others

Metallic and non-metallic elements do not occur in the organism in their free form; they are either ionized or else they form covalent bonds with other molecules. Our understanding of the essential nature of minerals for the organism is undergoing a rapid transformation. A mineral is considered to be essential if:

(1) a deficiency of the substance causes a functional disorder or some other form of reduced function;
(2) a supplementation with the mineral is of importance to development;
(3) deficiency symptoms are associated with decreased concentrations of a mineral in the blood or in other bodily fluids.

Our basic information about the necessity for different minerals must necessarily be based on animal experiments, because it is of course quite out of the question to conduct experiments in which a specific substance is excluded from the diet of humans in order to see if they will become ill. An exception is where deficiency symptoms which are associated with a specific mineral have already occurred in human beings, as was the case for selenium deficiency with Keshan's disease in China (see 5.3).

Although the quantities of minerals in the body are very tiny, their importance is correspondingly enormous. Minerals participate in a great number of enzyme and metabolic processes. Minerals can either reinforce or counteract each other in the organism, and the quantities in which they are found give no clue as to how vital they are for health. Tiny quantities of selenium are every bit as important as large quantities of calcium and magnesium. Minerals contribute to the building up of the tissues. Calcium, phosphorus, zinc, magnesium, silicon and fluorine are components of bones and teeth, while sulphur and selenium participate in certain amino acids (cysteine and methionine) which are the building blocks of hair, nails and skin. Iron and copper are essential ingredients in haemoglobin and myoglobin in the blood. A vitamin B12 molecule contains an atom of cobalt and iodine is vital to the functioning of the thyroid gland.

Minerals are the cornerstones of thousands of enzymes and chemical compounds. Sodium, potassium, calcium and phosphorus regulate the acid–alkali balance. In addition, sodium and potassium have an influence on the osmotic pressure in the tissues and on the elimination of various fluids from the organism. In this section we have only very briefly touched upon some of the properties of minerals for the organism. In the following sections, where we consider minerals one by one, we will consider in more detail which foodstuffs are good sources of minerals, what degree of bio-availability they have, and other properties.

2.   MINERALS AND TRACE ELEMENTS IN DIET

You can tell good news at any time, but bad news has to wait until the morning.

George Herbert (1593–1633)

In Finland, intense scientific research has been conducted into the mineral content of diets for many years. The studies referred to in this chapter describe some details of these results concerning Finnish diets.

In 1981, the results of a Finnish multicentre study led by Professor **Leena Räsänen** were published. This was an interview study which focused on the vitamin, mineral and trace element content of the diets of 3- and 12-year-old children. The results concerning minerals and trace elements are shown on page 152.

The most comprehensive mineral and trace element study to date in Finland was carried out in 1977–1978 by Professor **Pekka Koivistoinen** of the Institute of Food Science and Chemistry at Helsinki University.

On this occasion, the dietary content of 24 different minerals and trace elements plus 4 heavy metals was investigated. The quantities were measured in an average male diet consisting of 2800 daily calories. Possible quantities of left-overs were not taken into account, while those with a smaller daily energy intake (women, children, the elderly) would have a correspondingly lower intake of these substances. For example, the average intake for women is 30% less.

The results of the study are summarized in the table on page 153.

Whole grains were the source of more than 40% of the total intake of iron, copper, manganese and nickel in this study. White flour contains iron supplements which cover 20% of daily iron requirements (4 g). Whole grains contain only very small quantities of calcium, selenium, zinc and boron and cover less than 10% of the daily intake of these minerals.

More than 30% of the Finnish zinc and selenium intake comes from *meat*. On the other hand, meat contains virtually no calcium, manganese, nickel or boron. About 30% of selenium and 60% of mercury comes from fish.

Dairy products contain a great deal of calcium (more than 80% of intake comes from this source) and phosphorus (50% of intake). Dairy products also cover 30% of the requirements of potassium, sulphur, zinc and molybdenum. On the other hand, milk is low in iron, manganese, nickel, silicon and boron.

Minerals and trace elements	Average daily intake for Finnish children		
	3 years	12 years	
		Boys	Girls
Potassium, mg	2750.0	4140.0	3340.0
Calcium, mg	960.0	1342.0	1133.0
Magnesium, mg	225.0	370.0	291.0
Phosphorus, mg	1145.0	1841.0	1444.0
Iron, mg	9.1	17.8	13.4
Copper, mg	1.1	1.9	1.4
Manganese, mg	3.5	5.9	4.6
Zinc, mg	8.4	14.4	11.2
Chromium, μg	23.7	38.7	30.9
Fluorine, μg	230.0	420.0	330.0
Selenium, μg	11.4	20.1	13.8
Molybdenum, μg	76.0	115.0	90.0
Bromine, mg	1.9	2.6	2.2
Aluminium, mg	3.5	6.1	4.8
Silicon, mg	19.6	29.1	23.7
Boron, mg	1.1	1.4	1.2
Nickel, μg	88.0	132.0	96.0
Arsenic, μg	33.0	46.0	37.0

mg=milligrams, μg=micrograms.

Vegetables have a high content of boron (covering 60% of intake) and potassium (30% of intake). Various beverages provide a considerable proportion of the intakes of fluorine (40%, mostly from tea and salt), silicon (40%, mostly from beer) and aluminium.

These studies on the average Finnish intake of various minerals resemble fairly closely the few studies conducted in other western countries. There are, however, certain differences. Calcium intake is much higher than in other countries. Finnish table salt (sodium) consumption is also very high. Sodium and potassium are competing minerals, as are calcium and magnesium. A high sodium and calcium intake can therefore disturb the potassium–magnesium balance. The Finns probably suffer from a relative magnesium deficiency, at least in relationship to calcium, although total magnesium intake is within the recommended limits. Moreover, the Finnish diet is very high in phosphorus and low in chromium, fluorine and copper, compared to the diet in other countries.

3. SELEMIUM

The truth is tough. It will not break like a bubble at a touch; nay, you will kick it about all day like a football, and it will be round and full at evening.

O. W. Holmes (1809–1894)

Selenium has become the central topic of discussion in the debate on trace elements in

Minerals and trace elements	Average daily intake (Finnish male diet)
Sodium	17 g
Potassium	4500.0 mg
Calcium	1500.0 mg
Magnesium	440.0 mg
Phosphorus	2000.0 mg
Sulphur	1200.0 mg
Iron	19.0 mg
Copper	1.7 mg
Manganese	6.1 mg
Zinc	16.0 mg
Molybdenum	120.0 μg
Cobalt	13.0 μg
Nickel	130.0 μg
Chromium	29.0 μg
Fluorine	250.0 μg
Selenium	30.0 μg
Silicon	29.0 μg
Rubidium	5.6 mg
Aluminium	6.7 mg
Boron	1.7 mg
Bromine	4.2 mg
Mercury	5.7 μg
Arsenic	58.0 μg
Cadmium	13.0 μg
Lead	66.0 μg

g=grams, mg=milligrams, μg=micrograms

which some doctors have dismissed the mineral as humbug while others recommend it as a valuable component in preventive medicine, or even prescribe it as part of a course of treatment. Veterinary surgeons, on the other hand, have been using adjunct selenium therapy for many years.

Selenium is now generally recognized to be a trace mineral of great importance for human health. It is one of the most important antioxidants, which protect our cells from the harmful effects of rancidification. Selenium counteracts cancer and chromosome damage as well as increasing our resistance to viral and bacterial infections. It is because of the overwhelming significance of selenium that this section is one of the most comprehensive in the book.

Selenium is chemically similar to sulphur. **Jöns J. Berzelius,** a Swede, discovered selenium in 1817 while he was doing research into sulphur compounds. When selenium replaces sulphur in proteins, seleno-proteins such as seleno-methionine and seleno-cysteine are formed, which are components of the important cell enzyme, glutathione peroxidase. A molecule of glutathione peroxidase contains four atoms of selenium.

A short history of selenium
When **Marco Polo** undertook his voyages of exploration to Western China in the thirteenth century, he described an illness which affected horses, which was later discovered to be selenium poisoning.

Doctors attempted to use selenium in the treatment of cancer in Germany, France, Britain and the United States at the beginning of this century. The dosages used were, however, far too high and caused selenium poisoning, with the result that selenium became anathema to the medical profession.

In the 1930s selenium poisoning was also detected in livestock on the American prairies. This was caused by the high selenium content in cattle feed. Animal experiments have since revealed that large doses of selenium can be toxic for animals. In 1941 the first signs were detected that selenium might also be an important nutrient for living human organisms and in 1949 it was discovered that it could prevent liver tumours in rats.

In 1957 the first major breakthrough was achieved by Dr **Klaus Schwartz.** He was experimenting at that time with a diet of torula yeast, which caused liver necrosis in rats. He discovered that this condition could be prevented with the aid of kidney extract which later turned out to contain selenium. When this information was released to the rest of the medical research world, selenium became the object of the intense programme of research which continues to this day, more than 30 years after Klaus Schwartz's discovery.

The early research was concentrated on the relationship between selenium deficiencies and various animal diseases. The experiments revealed that selenium could prevent muscular dystrophy in calves, sheep, pigs and hens. It was also discovered that selenium was effective in the treatment of other livestock diseases and that in some cases vitamin E could complement, or even replace, selenium in these treatments. In 1961 New Zealand, which like Northern Europe and Scandinavia is one of the areas of the world with the lowest levels of selenium in soils, was the first country to make widespread use of selenium in order to prevent deficiency symptoms in domestic animals.

In Finland, selenium was employed in the treatment of muscle diseases in domestic animals for the first time in 1967, and two years later the veterinary authorities approved of the addition of selenium to animal feed. Since 1984 the Finns have added selenium to all fertilizers to increase the selenium content of agricultural products, and have thus indirectly increased the selenium intake of the human population.

In the early 1970s it was discovered that selenium prevented mutations (changes in the genes) but it was not until 1973 that researchers detected one of the enzymes through which this micronutrient functions. A glutathione peroxidase molecule contains four atoms of selenium and this enzyme is dependent on the supply of selenium for its activity in the blood and in the tissues. Glutathione peroxidase also plays an important part in the defence of the cells in that it fights peroxidation.

It was therefore possible to prove that selenium occupies a vital role in the prevention and treatment of a range of diseases in the field of veterinary medicine. Nevertheless, it took a long time before doctors were prepared to accept that selenium also has an important preventive function in human medicine.

The first information about selenium deficiency in human beings reached the West from China in the late 1970s. Researchers at Xian University had started experimenting with selenium supplementation on human beings in 1965, in connection with a heart muscle disease which affected especially young women and children. This disease has come to be called Keshan disease and is named after a province in China which is considered to be the area with the lowest selenium levels in the world. Many people died of the disease but with the aid of selenium supplementation it has more or less become extinct.

Since then, doctors in the Soviet Union and China have discovered another disease connected with selenium deficiency which can be prevented and treated with selenium supplementation. This is called Kaschin–Beck disease and affects the cartilage in the joints.

It has more recently been established in many other parts of the world that selenium deficiency increases the risk of heart diseases and cancer.

Sources of selenium and selenium requirements

Scandinavia and Northern Europe are amongst the areas of the world with the lowest selenium levels. During the glacial ages the selenium-rich earth layers were pushed southward, with the result that agricultural land contains only very small quantities of this vital micronutrient. This situation has been compounded by pollution, acid rain, the exhaustion of the earth and the intensive use of artificial fertilizers. In addition, the modicum of selenium which remains exists in a form which is not readily available to crops.

In the 1970s, selenium-deficient agricultural land represented a serious problem for the Finnish cattle industry and for other domestic animal industries. This problem has since been completely eradicated by the addition of selenium to cattle feed and fertilizers.

In general, foodstuffs in Scandinavia are very low in selenium. According to the Danish Food Authorities, the average daily selenium intake in Denmark in 1988 was 45 μg, the same figure as for the Swedes. Norwegians consume 50–70 μg per day while in Finland the figure is 90–100 μg because of the above mentioned precautions, involving the enrichment of animal feed and fertilizers. It is important that we emphasize that these are averages: some people get more and others significantly less than the average. As a comparison, in Keshan province the daily selenium intake was less than 10 μg per day.

The exact minimum requirements are not yet known. In the United States the figures have been based on experiences with animals and 50–200 μg has been adduced to be a 'safe and sufficient' level for daily intake. In Denmark the authorities have chosen a compromise between the two figures above and have set the recommended dietary allowance at 125 μg. On the other hand, an intake of 70 μg for healthy men and 50 μg for healthy women suffices to keep the kinetics in balance (i.e., the absorption corresponds to the excretion). However, this would not take into account the question of the prevention of disease.

Dr **Gerhard Schrauzer** from the University of California, like many selenium researchers, is of the opinion that selenium intake should be at the level of at least 200–300 μg per day, because lower intakes do not have the same prophylactic properties in relation to cancer. My own research confirms Schrauzer's views, since a daily intake of 200 μg of organic selenium is needed to control lipid peroxidation in humans.

The meagre selenium intake amongst the Scandinavian populations is reflected in their blood selenium levels. Danes and Swedes have the lowest recorded levels together with Egyptians and New Zealanders. These figures are only half as high as those recorded in Canada and the United States. People who eat a lot of fish generally have higher blood levels of selenium but the selenium in fish is not so easily available for the organism.

As we have mentioned, selenium has been added to artificial fertilizers in Finland in an attempt to increase the quantities of the trace mineral in grains. This has increased the average daily intake from 30 to 90 μg per day. The selenium content in Finnish human blood serum has risen, on average, from 75 to 100 μg l^{-1}. There are, however, still a considerable number of people whose serum selenium levels lie below 85 μg l^{-1} and this may have a significance for the increased risk of cardiovascular diseases and cancer.

Risk Groups

Certain groups in the population are especially at risk of a suboptimal selenium intake. These include:

(1) *Young people*   Many young people who have left home, or who live in student halls of residence, do not take very much trouble to ensure that they get a balanced diet. Finnish studies have revealed that 50% of students receive less than half of the average daily selenium intake. Autopsies on car crash victims have shown that the 15–30 age group have the lowest selenium blood levels.

(2) *Vegetarians*   Grains, plants and vegetables are so low in selenium that vegetarians often have very low selenium blood levels. In many cases a vegetarian's daily intake is as low as 10 μg per day. This is only a fraction more than what the victims of Keshan disease received in China.

(3) *The elderly*   Good selenium sources are meat and fish. Most elderly people eat smaller quantities of meat than are necessary in order to cover their minimum selenium needs. This is often the result of economic factors or because they have bad teeth or false teeth which make it difficult to chew.

(4) *Pregnant and nursing mothers*   Young women may develop selenium deficiency during pregnancy, largely because the foetus requires its share of the mother's selenium reserves. During breast feeding the mother loses selenium every day through the concentrations in breast milk. The deficit can rarely be made good by dietary adjustments, and selenium supplementation is needed.

(5) *Smokers*   A number of studies have shown that smokers generally have much lower selenium blood levels than non-smokers. Smokers either do not absorb

selenium adequately, or their needs are quite simply greater, because smoking and alcohol result in an increase in the formation of free radicals.

(6) *The chronically ill*   A number of gastrointestinal diseases increase the risk of selenium deficiency. These include malabsorption, celiac disease or gluten allergy and diseases or medications which produce loss of appetite, unvaried diets, diarrhoea and frequent vomiting. Lower blood selenium levels have been recorded in cancer patients as well as patients suffering from other chronic diseases, such as heart diseases, and in arthritis patients undergoing cortisone treatment.

The selenium metabolism

In food selenium occurs only in its organic form, whereas it is found in food supplementation products in both organic and inorganic form. Selenium supplements for domestic animals are usually inorganic while those designed for humans are usually organic, although inorganic selenium supplements are still to be found in health food shops.

Organic and inorganic selenium are absorbed in the body in different ways. Their bio-availability — i.e. the degree to which they are made use of in the tissues and organs — varies considerably. Most organic selenium compounds are almost totally absorbed (85–95%), whereas there are considerable individual variations in the absorption of inorganic selenium. Absorption percentages for the latter vary from 40 to 70%. Differences can also arise depending upon whether the inorganic selenium is in the form of selenate or selenite.

Organic selenium is stored in the tissues, while inorganic selenium is eliminated with the urine when the organism reaches saturation point.

Human blood contains about 0.1 mg of selenium per litre. 60% of the selenium in the blood is located in the red blood corpuscles while the other 40% is found in the serum. The selenium is transported with the blood to all the tissues in the body. The human adult body normally contains 10–15 mg of selenium (in selenium-poor areas like Scandinavia only 5–6 mg). Approximately half of the total quantity is to be found in the liver. In the tissues a proportion of the selenium is bound to the enzyme, glutathione peroxidase, while the rest binds itself to haemoglobin and other proteins. The discovery of these new selenium proteins is very new and their significance is not yet known. Similarly, it is known that selenium is a component of other enzymes which have not yet been thoroughly researched. The activity of the selenium enzyme, glutathione peroxidase, is proportional to the selenium blood values until this enzyme's activity level reaches the optimal.

Patients who are undergoing adjunct selenium treatment and people who take daily selenium supplements should have whole blood levels for this micronutrient of between 200 and 350 μg l^{-1}. When selenium values are at this level, selenium exerts an inhibitory effect of lipid peroxidation and other beneficial functions listed below.

The effects of selenium

Selenium has several different tasks in the body.

● It is an antioxidant, i.e. a substance which prevents rancidification (lipid peroxidation) of the cells and cell damage and which delays the pathological ageing process.

- It prevents blood clots by inhibiting platelet aggregation.
- It increases the effectiveness of the immune system and strengthens resistance to viral and bacterial infections.
- It inhibits chromosome damage, mutations and cancer.
- It counteracts adverse effects of heavy metals and other toxic substances in the body.

Selenium is an antioxidant

Selenium is a part of the body's defence system and protects the cells against the damage caused by oxygen free radicals. The selenium enzyme, glutathione peroxidase, breaks down hydrogen peroxides which occur in the cells and it would appear that it also breaks down fatty acid peroxides. Vitamin E complements the protective effects of selenium and participates in the destruction of free radicals. We can think in terms of vitamin E being the 'front-line' while selenium works 'behind the lines' — selenium destroys the peroxides which vitamin E is unable to deal with and the production of free radicals is minimized. Enzymes, such as catalase and superoxide dismutase (SOD) also participate in this defence system (see the section on copper: 5.12).

Selenium strengthens resistance

Selenium enhances the functioning of the T-lymphocytes and the macrophages (scavenger white blood corpuscles). The T-lymphocytes are programmed to recognize virus, bacteria and cancer cells so as to be able to produce antibodies against them more rapidly. These become so-called killer cells which are able to destroy cancer cells. The macrophages digest and destroy infection cells which have been invaded by viruses and bacteria. This process is described in more detail in the section on the immune system (see 1.6).

Selenium prevents infections and blood clots

Selenium's significance for the metabolism of prostaglandins is at present the object of active research. The beneficial prostaglandins are an important regulatory factor in the coagulation of the blood, and hence influence the incidence of arteriosclerosis and the clotting of blood platelets (thrombocytes). Consequently they also play a role in the prevention of thromboses. Some prostaglandins have harmful effects, they give rise to a form of inflammation response in the joints, for example. It would appear that selenium increases the production of beneficial prostaglandins and decreases the formation of harmful prostaglandins. Low serum levels of selenium (i.e. less than $85 \mu g \, l^{-1}$) increase the risk of a low level of 'the good cholesterol', HDL-cholesterol, and the risk of clotting of the thrombocytes. In addition, low selenium levels are associated with ECG abnormalities indicative of heart problems. It would appear that a low selenium intake is one of the contributory factors in myocardial infarction and in arteriosclerosis.

Selenium has anti-cancer properties

Animal experiments have shown that selenium deficiencies increase the risk of contracting cancer, that selenium supplementation prevents cancer and that selenium inhibits the transference of certain carcinogenic viruses from the mother to the foetus (Bittner's milk virus).

A whole wave of studies have been published to date in which mice, rats, hamsters and other experimental animals have been used to establish the importance of selenium in the fight against cancer. A food supplement of 1–5 parts per million has inhibited breast cancer, cancer of the large intestine, skin cancer, cancer of the liver, Ehrlichs ascites tumours, sarcomas and leukaemia. Selenium supplementation also makes it much more difficult to transfer a cancer from one experimental animal to the other.

Selenium also inhibits the transference of Bittner's milk virus, a carcinogenic virus in rats in which the foetus is infected by the pregnant rat. Other carcinogenic viruses have also been rendered harmless by selenium. Selenium supplementation has been effective in the prevention of many, but not all, forms of cancer in experimental animals.

Vitamin E and vitamin A reinforce the cancer-preventive properties of selenium. When selenium was used alone against experimentally induced cancers in experimental animals, the success rate was about 50% (the cancers were either chemically induced or transferred from other animals) but when selenium was combined with vitamin A and vitamin E the success rate was 90%.

Population studies

There is also a correlation between selenium deficiency and cancer in human beings. The incidence of cancer is more common in a subgroup of a population with low selenium blood levels. The differences were first seen in the 1960s in the United States, where there are both low-selenium and selenium-rich states and diets vary significantly in selenium content from state to state.

A large number of studies have shown that cancer patients who suffer from cancer of the breast, the kidneys, the bladder, the ovaries, the prostate, the rectum and the skin, together with leukaemia patients, have blood and tissue levels of selenium which are, on average, 30% lower than those of healthy controls. The interpretation of these findings is complicated by the fact that a seriously ill cancer patient is unlikely to eat the same food, or have as healthy an appetite, as someone who does not suffer from the disease. These factors presumably affect the selenium intake and that of other trace elements. Therefore the association between low selenium status and cancer does not prove a causal relation. This is a general problem in these case–control studies.

Several studies have been carried out in Finland, Sweden, Holland and the United States where blood tests have been taken from large groups of healthy individuals who were followed up over a period of many years. These studies suggest that those individuals who later contract cancer already have lower selenium values for years before they become ill, compared with those who do not contract the disease. Deficiencies of both vitamin E and selenium increased the risk of contracting cancer by a factor of 10, according to two major studies carried out in Finland. Potential cancer patients often belong to the lowest 20% in a scale of selenium levels.

If this is accompanied by deficiencies of other antioxidants, such as β-carotene, vitamin E, and vitamin A, then the probability of contracting cancer is further increased.

Selenium would appear to prevent cancer in at least four different ways:

- Selenium protects the cells from damage caused by oxygen free radicals. Free radicals form peroxides which speed up the promotion phase of cancers, i.e. the appearance of precursors to cancer (see also 3.10).
- Selenium decreases the mutagenenesis of carcinogens. This means that chemicals, viruses and radiation have a less carcinogenic effect.
- Selenium inhibits the reproduction of carcinogenic viruses.
- Selenium inhibits the division of cancer cells.

Of course selenium is not the only significant factor in the development of cancers. There are many other factors which must be taken into account, whether or not they arise simultaneously with selenium deficiency.

Here I have only considered the importance of selenium deficiency in the occurrence of cancer. Selenium has also been employed in the adjunct treatment of cancer. The idea is not entirely new. The German doctor, **August von Wassermann**, used and recommended selenium injections as a treatment for cancer as early as 1911. Other doctors followed suit, including **C. H. Walker** and **F. Klein** in America in 1915 and **Watson Williams** in Britain in 1922.

A Finnish immunologist, Dr **Thomas Tallberg** has used selenium as an adjunct therapy for cancer since 1970. Later several other doctors in Finland followed suit.

When selenium is used in the adjunct treatment of cancer the dosages involved are significantly higher than those used in prevention. For chemoprevention, 200–300 μg per day are sufficient, whereas cancer treatment may require higher dosages.

Selenium protects the heart

A heart muscle disease, 'mulberry heart' of cattle, which is caused by selenium deficiency and which can be cured with selenium, has been known of in the field of veterinary medicine for many years. The problem is a degeneration of the heart muscle, with a resulting insufficiency.

In the late 1970s researchers in the West heard about a disease which affected young women and children in Keshan province in China, symptoms of which resembled those caused by the aforementioned animal disease. Measurements of the selenium content of grains from the area in question revealed that levels there were only a quarter of what was normal in other areas.

Many children and young people died and the survivors were made invalids for life. The disease is now virtually non-existent after everyone in the area was given selenium supplementation.

When the news about Keshan's disease spread to the rest of the medical world, researchers began to take a closer interest in the relationship between selenium and diseases involving degeneration of the heart muscle.

A number of incidences of the same sort of disease have been reported from elsewhere in the world but in all these cases it has been associated with patients who were already very ill and who were being fed via a drip (total parenteral nutrition).

Selenium and heart diseases

In 1982 a Finnish study showed that the risk of contracting heart disease was increased by a factor of between 2 and 7 over a period of 7 years, when serum levels of selenium were low (less than 45 μg l^{-1}). This report by Dr **Jukka Salonen** gave an impetus to the research in selenium and heart diseases in Europe. Dr Salonen's further research indicated that serum selenium levels below 85 μg l^{-1} are associated with risk factors of coronary heart disease, i.e. low HDL-cholesterol, increased platelet aggregation and abnormalities in the exercise ECG.

Selenium levels in serum also associate with the content of an essential fatty acid, eicosapentanoic acid (EPA), which has been shown to protect the heart. These two protective factors, EPA and selenium, both occur naturally in fish, allowing us to draw the conclusion that a diet which is rich in fish is good for the heart.

Patients with acute myocardial infarction are often deficient in selenium and in vitamin B6. Other antioxidants, including the trace mineral zinc and vitamins A, C and E are able to protect the heart against ischaemia and against reperfusion injury. A combination of water-soluble and fat-soluble antioxidants can provide the best protection to the heart. A recent Dutch study by Dr **Frans Kok** showed that selenium deficiency often precedes a myocardial infarction. A British study published in 1989 in the *British Medical Journal* confirmed that patients with heart infarction and arteriosclerosis indeed have much rancid fat (lipid peroxides) in their blood, and this phenomenon is of etiological importance. Animal experiments have so far supported this theory. Animals with heart disease get on better and produce fewer rancidified fatty acids when they are given selenium and vitamin E. In addition, the animals' hearts are better able to tolerate short-term oxygen deficiency than those animals which have not been given supplements. The size of myocardial infarction can be reduced by antioxidants.

A whole series of scientific reports have illustrated how antioxidants also prevent reperfusion injury and arrythmia in animals and in heart patients. Judging by the available evidence, heart patients ought to be given supplements of selenium and other antioxidants. I have already described some of these case histories, where antioxidant treatment proved to be beneficial (see 2.4).

Selenium and chronic illnesses

Certain neurological children's diseases are also connected with levels of selenium intake. Intensive research has been carried out on juvenile seroid lipofuscinosis, Duchenne's muscular dystrophy and Becker's disease, against which adjunct antioxidant treatment (Westermarck's formula) has produced a series of encouraging results.

According to a new study, carried out at Roskilde University in Denmark, there is also a connection between the deterioration in the condition of patients suffering from disseminated multiple sclerosis and lipid peroxidation which in turn associates with selenium deficiency. Professor **Jørgen Clausen** has provided his sclerosis patients with selenium supplementation, with positive results.

Free radicals and lipid peroxides are formed whenever an inflammation response arises. According to my personal experience, selenium in combination with other antioxidants can therefore alleviate rheumatic pains and stiffness in the joints in various rheumatic and arthritic complaints. Other experimental treatments have shown that in some instances selenium supplementation also has a beneficial effect on the neurological disease, myasthenia gravis.

It can also be worthwhile to give selenium and other antioxidant supplements in combination with essential fatty acids for various allergies, such as asthma, hay fever and eczema. I also recommend selenium and other antioxidant supplements as a prophylactic for diabetic eye complications and for injuries to nerves and blood vessels because of an increased lipid peroxidation in diabetes.

Selenium deficiency has been detected in a number of AIDS patients and experimental treatments are now being carried out with selenium supplementation in this field.

Selenium as an antidote to heavy metals and cytostatics
It has been discovered that selenium inhibits the harmful effects of heavy metals, such as arsenic, mercury, cadmium and lead, in experiments with animals. The mechanism of this phenomenon is not yet fully understood (see also 1.5).

Selenium also has a protective effect against other toxic substances, the cytotoxic drugs used in chemotherapy for example. Adriamycin, a commonly used medicine against ovarial cancer, increases the formation of free radicals and has a seriously toxic effect on the heart and the liver. It has been shown at Oulu University, Finland, that when selenium is administered together with adriamycin, the adverse side-effects are reduced.

Preparations of selenium
It is very difficult to manufacture a good organic preparation of selenium which can be absorbed and utilized effectively by the body, and there are considerable differences in the various products on the market. Professor **Jørgen Clausen** at Roskilde University, Denmark, has carried out a comparative study of the bio-availability of ten different preparations of selenium. The best preparation in his test was the antioxidant combination, Bio-Selenium, which also contains zinc and vitamins A, B6, C, and E. This incorporates selenium in an organic form, known as L-seleno-methionine. Inorganic selenium is much less effective, and consequently, the daily dose, in order to give the equal effect, must be much higher as compared to L-seleno-methionine.

The difference between organic and inorganic selenium is illustrated in a study involving breast-feeding mothers, which was carried out at a children's clinic in Helsinki. It was discovered that organic selenium was absorbed by the mother and secreted in satisfactory quantities into breast milk. Inorganic selenium, on the other hand was scarcely registered at all in breast milk. The selenium blood values of the infants were in close proportion to the secreted quantities in breast milk.

4. CHROMIUM

It was only late in life that I discovered how easy it is to say 'I don't know'.
                                                    Somerset Maugham (1874–1965)

Chromium is a much-discussed trace element. Chromium intake in the Scandinavian

countries lies consistently below the Recommended Dietary Allowances (RDA), and the Finnish chromium intake is probably the lowest in the whole world.

It is still too early to determine who really needs daily supplements of chromium — is it only necessary for diabetics and heart patients, or should everybody take them? As things stand, the individual reader must decide whether or not he or she requires extra chromium. In order to be able to take this decision, the prerequisite is adequate information, which I hope to be able to provide in this chapter.

Chromium requirements and metabolism

Requirements for chromium are not yet known. According to American estimates, a daily intake of 50–200 microgrammes is a safe and adequate dosage. Our actual intake, however, is considerably lower than these figures. In Finland, for example, the average daily intake is only $30\,\mu g$.

Experts are still in disagreement about who needs chromium supplementation. It now appears, in any case, that supplementation is required for the elderly, for diabetics and for people with heart and cardiovascular problems, since these groups are particularly exposed to the risks of chromium deficiency.

The best dietary sources of chromium are meat, whole grains, cheese and herbs. Brewers' yeast also contains chromium.

Chromium is found in both organic and inorganic form, but inorganic chromium (chromium III chloride) is very poorly absorbed in the organism. While only about one half of a per cent of inorganic chromium is absorbed, it now appears that up to 25% of organic chromium can be assimilated.

Chromium absorption can be inhibited by iron, manganese, calcium, zinc and titanium. In the blood, chromium is bound to the protein, transferrin, which also binds iron. Chromium is later released from the blood and is then stored in the liver, the spleen and the bone marrow. At this stage chromium is associated with vitamin B3.

Chromium is almost totally excreted with urine, while only a tiny quantity leaves the body with the faeces or with sweat. The best method of measuring the chromium balance in the organism is to observe the quantities excreted with the urine. Chromium content of serum and whole blood is minute, and measurements of these values can not be used to evaluate the chromium balance or to determine the effectiveness of chromium supplementation.

The effects of chromium

Chromium is an essential trace element, i.e. it is necessary for life. Chromium deficiency in experimental animals reduces protein production, shortens the life-span, leads to eye damage and affects fertility. It has also been discovered that chromium deficiencies in rats and monkeys lead to decreased glucose tolerance, which can later be normalized with chromium supplementation. Recent studies have indicated that this may also be the case for human beings.

It has been postulated that chromium in the organism is a component of a molecule, known as GTF (glucose tolerance factor), which is responsible for establishing a balance in the sugar metabolism. Organic chromium yeast may contain large quantities of this molecule, although the molecule has not yet been discovered.

According to a recent study, conducted by Professor **Jórgen Clausen**, it would appear that chromium is able to prevent the symptoms of low blood sugar, also known as hypoglycaemia. A number of other studies have pointed to a link between chromium and blood sugar balance, but most of these studies have lacked adequate control groups. Two other controlled studies have confirmed Professor Clausen's findings, but other similar studies have produced conflicting results. This may possibly be a consequence of the application of different preparations and different dosages of chromium.

For several years, researchers have suspected that chromium deficiency may be a contributory factor in arteriosclerosis. The chromium values in the aorta of arteriosclerosis patients are very low, according to the *post mortem* findings on the victims of heart attacks and traffic accidents.

Chromium values fall correspondingly with increasing age in countries where arteriosclerosis is rife, in contrast to countries where this disease is rare. In Finland, the inhabitants of the eastern part of the country are especially prone to heart diseases, and chromium and selenium deficiencies are suspected as a possible factor in this tendency.

A number of studies have suggested that organic chromium (chromium yeast) may be able to reduce the lipid (fat) levels in the blood. There is, however, a lack of cross-over-double-blind research in this field, so this question must remain unanswered for the interim.

Chromium therapy
There are, as yet, no commonly accepted grounds for the use of chromium in the treatment and prevention of illness. I have prescribed chromium on many occasions, often together with zinc and selenium, mainly as adjunct therapy for patients with diabetes, hypoglycaemia, coronary artery disease and arrhythmia. Several of my diabetic patients have experienced reduced sugar levels in the blood combined with a decreased insulin requirement. Some patients were even able to stop taking insulin altogether. Moreover, many of my patients reported reduced symptoms and an improvement in their general health. It is difficult to ascertain whether or not these effects have really been caused by chromium supplementation, however.

A new Korean study revealed that 83 out of 100 diabetes patients experienced a reduced dependency on insulin and an improvement in their general condition after supplementation with chromium. The 17 patients who experienced no improvement either had a total failure of the pancreas or were given medicine which counteracted chromium. For chromium to be able to regulate the sugar balance requires the presence of just a tiny quantity of insulin, so the chromium can replace the insulin which is missing. This is not possible if the pancreas's production of insulin has ceased altogether.

Chromium may also have another beneficial effect, as we have referred to above, in the regulation of the sugar balance for patients suffering from symptomatic hypoglycaemia, which results in symptoms of headaches and general discomfort.

5.   FLUORINE

Lack of information breeds suspicion

Francis Bacon (1561–1626)

The adult human organism contains 3–7 mg of fluorine. Fluorine in diet and human

tissue occurs in the form of fluoride. In experimental animals, fluoride deficiency causes limited growth, hair loss and disorders in the development of teeth. This last phenomenon has been cited as an argument for classifying fluorine as an essential mineral, although not all researchers agree. The debate has been particularly heated on the question of whether or not fluoride should be added to drinking water in an attempt to prevent dental decay. Opponents of fluoridation of water supplies emphasize the possibility that fluoride accumulations in the organism may be harmful.

Fluorine requirements and metabolism

In the United States the RDA for fluorine is 1.5–4.0 mg for healthy adults. Some researchers maintain that 4 mg should be the absolute maximum dosage. Natural fluorine intake varies in accordance with fluorine contents in soils.

Natural sources of fluorine are fish, in particular herring, sardines and mackerel. Tea also contains fluorine and it may be that tea is the major fluorine source for many people. The 'officially' recommended maximum dosage is 20–80 mg per day, while the fatal dose for adults is 2500–5000 mg of sodium fluoride (2.5–5 g — or a teaspoonful).

Dietary fluoride is rapidly and totally absorbed in the organism. Fluoride absorption can be affected by other dietary minerals and fats: absorption is reduced by aluminium, calcium and sodium chloride (table salt), while fats increase absorption.

Fluoride is transported via the blood to the bones, the teeth, the aorta, the kidneys and other tissues. Fluorine content in the bones increases with age if the water supply is fluoridized, while levels in softer tissues remain constant or only increase very slightly.

Fluoride is eliminated with the urine, which explains why fluorine loss decreases in patients with kidney failure. Urine content of fluoride reflects the level of intake but can also be strongly affected by previous intake.

Effects of fluoride

The most important function of fluoride is to prevent caries in the teeth. It strengthens the enamel and thus protects the teeth from the attacks of bacteria and acids. In some places, France for example, it is believed that fluoride can also strengthen the bones and that supplementation with this mineral can prevent and slow down osteoporosis.

This effect is a matter of dispute. A Danish study, which compared the use of fluorine with the use of hormones in the prevention of osteoporosis after the menopause, came to the conclusion that fluorine has no such preventive effect. In Finland, two separate studies reached conflicting conclusions. According to one of these studies, the risk of breaking the femoral collum was increased when fluoride intake was too low and when it was too high. The other study showed no effects of fluoride intake.

It has, however, been established that fluoride, together with vitamin D and calcium, increases the bone mass in cases of osteoporosis. The newly formed bone mass does indeed harden, but the resultant bone tissue cannot be compared to normal bones.

Fluoride has also been associated with both increased and decreased incidence of cardiovascular diseases. The lack of any uniform picture may be due to different methods of investigation or because various other parameters, such as the hardness of the water, dietary habits or the intake of vitamins, minerals and pollutants in the water supply, have not been taken into consideration.

Fluoride therapy
The use of fluoride for the prevention of dental caries is generally accepted. There are seldom grounds for measuring the plasma levels of fluoride, because intake is more or less dependent upon the content in drinking water.

In France, sodium fluoride is registered as a medicine for the treatment of osteoporosis.

6. SODIUM
Salt and thickeners are sweet for the poor.

Finnish proverb

The adult human body contains about 100 g of sodium, half of which is located in the cells, primarily in the bones. The remainder is concentrated in tissue fluids which surround the cells. Table salt consists of sodium chloride, and the organism receives its daily sodium requirement from this — and usually a great deal more. Salt is a common additive in foodstuffs and as a consequence it is difficult to affect or control salt intake.

Over-consumption of salt increases the risk of high blood pressure in people who are predisposed to hypertension. Healthy salt habits have to be learned in childhood because children often develop a taste for salty food from their parents.

Sodium requirements and metabolism
The human sodium requirement is about 3 g per day. However, most people consume between 5 and 15 g, and some even more. In other words, the body's requirement is exceeded many times. Half of our salt consumption stems from preprocessed foods, 40% is added while we prepare food or at the table. Only 12% of our daily consumption comes from the natural sodium content of foodstuffs.

High blood levels of sodium can be caused by over-consumption of table salt, acetyl salicylate-containing headache tablets or soda-containing mineral waters or soft drinks. Calcium or potassium loss can also contribute to a relatively high sodium level in the blood.

Most sodium is absorbed in the small intestine, and a small quantity is absorbed from the stomach. It is filtered from the blood by the kidneys and re-released into the organism according to needs. The kidneys are therefore responsible for sodium regulation (at the expense of potassium; see next section). Over 90% of sodium intake is eliminated with the urine.

The effects of sodium
Sodium regulates the electrolyte and acid–alkali balances, the conductive capacity of the nerves, muscle contractions, and the production of adrenaline and amino acids. An excessive sodium intake can raise blood pressure, at least for people who already have a tendency to this problem. Salt can also aggravate or maintain an already high blood pressure. High sodium intake can also lead to fluid build-up (oedema) and contribute to migraines.

Sodium deficiency is extremely rare. A malfunction in the kidneys can lead to disorders in the recycling of sodium, which can in turn cause oedema or lead to sodium or potassium deficiencies. Sodium deficiency can also arise as a result of heavy sweat secretion. If daily fluid requirements exceed 4 litres, then salt should be added to liquids (as was done in the past for agricultural workers during the harvest season).

Symptoms of sodium deficiency are general indisposition, dizziness, weakness in the muscles, loss of weight, respiratory problems and mild fever.

Sodium therapy
There are no grounds for treatment with sodium outside of the hospital. On the contrary, there is every reason to warn against excessive salt consumption. We should try to add as little salt as possible to food, because when the taste for salty food has been developed, we tend to continue to increase, rather than decrease, our salt consumption. This habit makes it increasingly difficult to do without salt in any situation.

It is well worth teaching healthy salt habits to children. There is no danger of a deficiency, and it could be that the child will benefit from this in later years, if he or she is forced to cut down on salt intake, because of high blood pressure, for example. I usually recommend herbs and spices or herbal salt which gives a taste to food, for this eventuality. This contains far less sodium chloride than normal salt. One can also use mineral salt, Pan-Salt, for example, where a proportion of the sodium is replaced by potassium, which is much healthier for the body.

7. POTASSIUM

The more one knows, the more sceptical one becomes.
                                        Michel de Montaigne (1533–1592)

Potassium is one of the minerals of which there are the largest quantities in the human organism: 115–150 g. Almost all of this — 98% — is located in the cells. Diuretic medicines can increase potassium (and magnesium) loss to such an extent that potassium deficiency eventually arises. Even healthy people can develop potassium deficiency, which can lead to fatigue, weakness, cardiac arrhythmia and sleeplessness. Potassium plays a vital role in the nervous system, the muscles and the heart.

Potassium requirements and metabolism
The average male diet consisting of 2800 kcal contains about 3–4 g of potassium and most people have a potassium intake of between 2 and 4 g per day on average. A

quarter of this comes from dairy products, a quarter from plants and vegetables, 15% from grains and 15% from coffee! There are high concentrations of potassium in citrus fruits and juices, green vegetables, bananas and potatoes.

Potassium supplementation, e.g. in the form of mineral salt, can reduce the risk of developing high blood pressure and may be able to reduce already high blood pressure. Dietary potassium does not represent any danger to health but I would certainly not recommend coffee as a source of extra potassium.

Potassium is absorbed in the intestines and is eliminated via the urine. If potassium intake is less than the amount lost, the potassium content within the cells and in the serum will fall. The cells then begin to use protons (H^+) instead of potassium (K^+) and this leads to acidosis, i.e. acidity.

Potassium loss (elimination) can be increased by many factors, including over-consumption of coffee, sugar and alcohol. Diuretic medicines can also increase potassium loss and treatment with these medicines should always be followed up by measurements of the potassium balance. There are diuretics which are potassium-retaining or to which potassium is added. Normally, when potassium elimination increases it is accompanied by magnesium loss but this is rarely detected, because measurements of magnesium blood values are seldom conducted.

The effects of potassium
Potassium and sodium are partners in the mineral world. This is because they have the opposite effects. Potassium is required for the normal functioning of the nerves and muscles, the sugar metabolism, acid–base balance and the oxygen metabolism in the brain. The heart also needs potassium. The correct potassium balance is necessary in the heart to avoid arrhythmia and the damage that ensues from such abnormal rhythms. In addition, potassium participates in a number of enzyme systems and in the protein metabolism.

Potassium deficiency can express itself as fatigue, muscle weakness, fluid accumulations, constipation and disorders of the nervous system and the kidneys. A deficiency of this mineral can also increase risk of digitalis poisoning for those undergoing treatment with this medicine because of a heart insufficiency. A high potassium content in the blood, on the other hand, can be caused by excessive fluid loss due, for example, to repeated diarrhoea or vomiting. High potassium blood levels can also produce symptoms of weakness, drowsiness, arrhythmia and sensations of disorientation.

Excessive and deficient intakes of potassium can therefore both lead to metabolic disturbances. Such occasional disturbances are common amongst diabetics. More chronic disorders are seen in cancer, heart, liver and kidney patients. Hyperthyrosis (hyperactive thyroid gland) can also result in potassium deficiency.

The kidneys do not regulate the potassium balance as efficiently as they do the sodium balance. Patients with kidney failure often show increased potassium values because of ineffective elimination of this mineral, and these effects can be observed, as a rule, as typical changes on an ECG (electrocardiogram).

Very high potassium intake may also reduce the effect of anticoagulant medicines (blood-thinners), and sometimes it becomes necessary to increase the dosages of anticoagulants when the patient increases his or her consumption of potassium-containing vegetables in the summer time.

Potassium therapy

Potassium deficiency, caused for instance by diuretics, should be treated with potassium supplements of 2–4 g per day (after potassium blood levels have been checked). Alternatively, a preventive dose of 1–2 g can be administered.

Potassium tablets can give rise to side-effects in the form of stomach trouble, constipation or irritation of the mucous membranes. People with a tendency for stomach ulcers can develop ulcers in the duodenum from these tablets. There are tablets which release potassium more slowly and which give fewer side-effects than the more rapidly dissolving tablets. Potassium tablets should be taken with plenty of water, otherwise the tablet may get stuck in the oesophagus and this can cause severe chest pains.

Fluid accumulations (oedema) can be reduced with potassium therapy. In the event of oedema in connection with pregnancy, the causes of the problem should be investigated before any treatment is administered. However, in cases of harmless swellings, potassium therapy is, in my opinion, often a better alternative than diuretics, which do not attack the source of the fluid accumulations and which may also lead to potassium deficiencies.

Although coffee is rich in potassium (45 mg per cup), a high consumption of coffee can lead to potassium deficiency and fatigue, the opposite of the desired effect for the coffee drinker. Coffee is, in fact, a diuretic, and potassium loss is often greater than intake when coffee consumption is high.

Potassium and magnesium deficiencies produce similar symptoms, and deficiency of one of them usually implies a deficiency of the other. It is therefore advisable to take these two minerals together. Potassium supplementation increases potassium concentrations in the blood but not in the cells; potassium deficiency in the cells can only be corrected with a combined supplementation of potassium and magnesium, a fact which both doctor and patient should be aware of. Calcium can partially correct symptoms caused by potassium deficiency.

Patients with kidney failure have a very poor tolerance for potassium supplementation.

8.   CALCIUM (CHALK)

The tragedy of ageing is not that we get old, but that we are no longer young.
                                        Oscar Wilde (1854–1900)

Calcium is the mineral which occurs in our bodies in the largest quantities — around 1200 g for an adult human, 99% of which is concentrated in the bones and teeth. Calcium supplementation is particularly important for babies and children while the bones and teeth are growing and developing.

Osteoporosis affects most women after the menopause. In industrialized countries calcium intake is higher than anywhere else in the world and actual calcium deficiencies are extremely rare. Nevertheless, osteoporosis is becoming increasingly common and certain population groups are exposed to a risk of calcium or magnesium deficiencies. Calcium and magnesium are competing minerals, and must be present in the organism in the correct proportions to each other.

Calcium requirements and metabolism

Calcium levels in human beings are dependent upon absorption through the intestinal walls. The degree of absorption is dependent on the presence of vitamin D, although a number of other factors, such as stomach acids, intake of dietary proteins and fibres and other solids, also play a role. Calcium intake amongst the elderly usually falls considerably. Long-term use of laxatives also reduces calcium absorption.

The RDA for calcium is 0.8 g for adults and slightly more for children. There are currently plans to increase the recommended daily dosage in order to prevent osteoporosis.

The best source of calcium is milk, which contains about one gram of calcium per litre. Cheese and other dairy products are also excellent sources of calcium.

The presence of calcium is also required for the absorption of vitamin B12 through the intestinal walls.

Every day, about 700 mg of calcium is transported to the bones from the blood, while equal quantities travel in the opposite direction. The skeleton is, in fact, living tissue with an active metabolism. Calcium is transported out of the body with the excreta, although insignificant quantities are also lost with the sweat. If blood levels of calcium are very high (more than $70 \, \text{mg/l}^{-1}$), the excess is eliminated with the urine.

The effects of calcium

Calcium participates in the regulation of nerve and muscle functions, in the production of hormones, in the maintenance of the fluid balance, in the activity of the heart, in the coagulation of the blood and in the secretion of milk. The inner and outer calcium balance in the cells is also of great importance. Calcium influx into the cell can lead to cell damage or even cell death.

Inadequate intake of calcium in children causes weakened development of teeth and bone mass, which, later in life, may lead to osteoporosis. Serum calcium levels are regulated by a hormone which is secreted from the para-thyroid glands. If too much of this hormone is secreted, because of a tumour for example, calcium content in serum rises.

Calcium and magnesium are reciprocal antagonists, i.e. they counteract one another. Magnesium supplementation can reduce high blood levels of calcium. On the other hand, calcium can correct possible potassium deficiencies. In order to correct a calcium deficiency, it is not enough to give only a supplement of calcium. Vitamin D is also required, because this vitamin is responsible for the absorption of calcium and for transporting it further to the bones.

Calcium therapy

Inadequate calcium intake weakens the bone mass, the teeth and the hair, and may also lead to allergic reactions (since calcium has anti-allergic activity).

Of all the disturbances caused by disorders in calcium metabolism, the most widespread — and the most difficult to treat — is post-menopausal osteoporosis. Attempts have been made to treat this problem, which occurs most commonly in women, with calcium alone, with calcium and vitamin D, and with combinations of calcium and sex hormones, oestrogen, fluorine and magnesium. Calcium supplements, however, have proven inefficient in this respect and recently it has been discovered that calcium is not assimilated in the bones on its own. This requires simultaneous supplies of silicon and magnesium.

Oestrogen therapy is based on the fact that oestrogen protects against osteoporosis before the menopause, but this is not always the case after the change of life. The purpose of fluoride therapy in this context is to facilitate the growth of bone tissue and avoid fractures, but the results of this treatment have not been convincing. Various medicines consisting of calcium-regulating hormones have also been tried, and the best results so far have been obtained with calcitonin, which is produced in the thyroid gland.

As is so often the case, it would appear that for osteoporosis, prevention is better than cure. This also provides the background for the American authorities' plan to increase the RDA for calcium. They hope to strengthen the development of bones in this way, as the best preparation for the inevitable decalcification which takes place after the menopause.

Calcium can also be of value in the treatment of allergies, for example, allergies to medicines. In such cases calcium can be administered in the form of injections or soluble tablets.

Mineral preparations of calcium may also be of value in various problems in the joints. My experience indicates that the best prognosis is obtained with a combination of calcium and magnesium, which also strengthens the hair and nails.

9.   MAGNESIUM

Be prepared for everything, answer to nothing, and do not be surprised by anything.

Arkhilokhos (670–620 BC)

There are 20–28 g of magnesium in the adult human body, of which 99% is located within the cells, where it regulates a number of enzyme systems. The greater part of our magnesium is concentrated in the bones (60%) and the muscles (20%), while only 1% is to be found in the fluids between the cells, in serum, for example. The small quantities of this mineral in serum mean that accurate measurements of magnesium levels in the tissues cannot be ascertained from serum values. Magnesium has many important functions in the organism; it improves resistance and protects the cells against a range of diseases. Recent observations have indicated that breast cancer patients have reduced magnesium levels in the red blood corpuscles. In Scandinavia, magnesium intake is very low, compared to the relatively high intake of calcium.

Magnesium requirements and metabolism

The American RDA for magnesium is 300–450 mg per day. The average Scandinavian diet can barely meet this requirement, and many researchers are of the opinion that magnesium intake should, in fact, be considerably higher than it is at present. At the beginning of this century, daily intake of this mineral was in the region of 1200 g.

Good sources of magnesium are plants, grains, meat and fish. Even a very unvaried diet should therefore not lead to serious deficiencies of magnesium, but this is offset by the fact that absorption capacity for this mineral varies considerably.

Many researchers deny that magnesium deficiencies of any kind occur in Scandinavia, while others maintain the opposite. I subscribe to the latter view after having observed that adjunct magnesium therapy has been associated with an improvement in the condition of many of my heart and allergy patients. Magnesium deficiencies amongst heart patients may be connected with the diuretic effects of anti-hypertensive drugs.

The absorption of magnesium from the intestines depends upon many factors. Generally, 30–40% of intake is absorbed, and when intake is reduced, absorption percentages increase. Not all the preconditions for the optimal absorption of magnesium are fully known as yet. A high intake of dietary calcium, phosphorus and proteins reduces magnesium absorption, and a number of other factors in diets also reduce absorption of this mineral, including phytate, saturated animal fats and fibres.

The magnesium balance in the organism is determined, to a considerable extent, by the degree to which magnesium is eliminated with the urine. Magnesium loss is usually about 100–150 mg per day and if serum magnesium values increase, the level of magnesium loss increases correspondingly. Calcium and sodium loss and some hormones also have an influence on the elimination of magnesium with the urine.

The effects of magnesium
The cells need magnesium

Magnesium is an essential mineral for cell functions and it occupies a key role in all reactions with phosphates.

The cells also require magnesium for cell division and enzyme production, which in turn regulates the protein, carbohydrate, lipid, nucleic acid and nucleotide metabolisms.

In the event of magnesium deficiency the permeability of the cell membranes is affected, causing potassium and magnesium to be forced out, while sodium and calcium force their way in. The influx of sodium releases calcium from the mitochondria and this gives rise to a substance called AMP, which causes the cell membranes to become even more permeable. This vicious circle eventually results in cell death.

The heart and muscles
Normal muscle function requires the presence of magnesium, which is why a deficiency of this mineral leads to muscle weakness and fatigue. Furthermore, magnesium is of importance for the heart and the entire circulatory system as magnesium prevents the blood pressure from rising. It now appears that heart patients are often deficient in magnesium and a correction of this deficiency may often lead to an alleviation of symptoms. A number of studies indicate that magnesium supplementation also reduces the risk of myocardial infarction, and specifically, the occurrence of a re-infarction later in life.

A *post mortem* survey, conducted on 1000 individuals at the turn of the century in Finland, revealed that not one of them had died of heart infarction. At this time, magnesium intake was three times as high as it is today, just as dietary intake of many other minerals, which we now know to be important for the heart, was higher.

Magnesium deficiency can also give rise to cardiac arrhythmia and increase the sensitivity of the heart muscle to digoxin. Serious long-term magnesium deficiencies may also cause hardening of the aorta and the small blood vessels in the arms, legs and kidneys. Controlled studies from Sweden have also suggested that magnesium may help in the prevention of kidney stones.

Strengthening resistance

Magnesium appears to improve the functioning of both the cellular and the antibody-mediated immune defence. The levels of antibodies (immunoglobulins) decrease in experimental animals (mice, rats and hamsters) by up to 60% when the supply of magnesium is significantly reduced. This decrease is greatest for IgG, but the levels of IgA, IgM and IgE also fall.

The reason for this fall in immunoglobulin production may be due to an inability on the part of the B-lymphocytes to develop into antibody-producing plasma cells. It may be that magnesium has other immunological properties which are mediated through hormone-like substances, in the reactions between antibodies and macrophages, for example.

The defence system, which is regulated by the T-lymphocytes, requires both magnesium and calcium. The T-lymphocytes are transformed into 'killer cells', amongst other things, which are able to attack cancer cells.

There is a direct correlation between magnesium deficiencies in rats and reduced immune defence against allergic reactions and cancers, in particular leukaemia and lymphomas. To date, no major population studies have been conducted on the possible relationship between magnesium deficiency and disorders in the immune system.

However, the medical literature abounds with descriptions of individual cases where magnesium deficiency has been associated with asthma and other allergic responses. Magnesium supplementation has led to reduced histamine blood levels and contributed to improvements in the condition of the patients in question, i.e. magnesium has had an antihistamine effect. I have observed similar effects in many of my asthma and atopic eczema patients. These patients often have very low magnesium levels in whole blood and when they receive magnesium supplements these values rise and symptoms are alleviated or disappear altogether.

The physical and psychological symptoms of undefined fatigue are often accompanied by insufficient magnesium intake.

Magnesium deficiency

Magnesium deficiencies in humans can have many causes. It can be as a result of an insufficient dietary intake, a reduced capacity for absorption or an excessive loss of magnesium with the urine. A number of diseases increase magnesium loss, including diabetes, kidney diseases and certain hormonal disorders. Magnesium loss with the urine is also increased by alcohol and by various diuretic medicines.

Magnesium deficiency is often accompanied by potassium deficiency, but regrettably, routine blood measurements of magnesium levels are rarely undertaken. This is mainly because serum measurements are not adequate for the detection of magnesium deficiency — this requires measurements of magnesium in the cells, platelets or whole blood. The best results, however, are obtained from the cumbersome measurements of the magnesium loss in the urine over a 24-hour period.

Magnesium deficiency can develop gradually without the appearance of readily detectable symptoms. Over the long term, however, fatigue, hormonal disorders, muscle weakness, trembling and possibly muscle cramps (restless leg) may occur. A serious magnesium deficiency, which may give rise to arrhythmia, can occasionally be detected on an ECG (electrocardiogram).

The relationship between heart diseases and magnesium deficiency has been the object of intensive research since the 1960s, when population studies revealed that hard water is associated with a reduced incidence of heart disease. The hardness of drinking water is dependent upon the amounts of calcium and magnesium in the supply, and it may be that it is magnesium which is responsible for the heart-protective properties of hard drinking water. There is still no conclusive evidence for this effect, but areas of low magnesium intake, Eastern Finland for example, show a correlation with a high incidence of heart disease. The magnesium content of hard water is higher than that of soft water.

The quantities of magnesium in the heart muscles of those who have died of heart attacks are lower than those of the victims of fatal traffic accidents, according to *post mortem* results. This observation may indicate that the risk of a heart attack increases with magnesium deficiency, although it could also be an indication that magnesium is lost from the heart muscle during the death throes. The implication here is that magnesium deficiency might just as easily be an effect, rather than a cause, of death. New and comprehensive population studies are therefore needed in order to shed light on this question.

A recent Swedish study has indicated a possible connection between breast cancer and blood levels of magnesium. Low magnesium values in the red blood corpuscles of breast cancer patients were detected in this study. Nevertheless, it is still too early to draw any conclusions about the extent to which magnesium may or may not be applied to the prevention or treatment of cancer.

At the Twenty-second Trace Mineral Congress in the United States in 1988, several magnesium researchers reported correlations between magnesium deficiency and cardiovascular diseases, high blood pressure and complications during pregnancy. Many researchers believe that the average magnesium intake in the USA is only 40% of the RDA and that over 80% of the population get less than the recommended daily dosage. Groups who are particularly at risk are excessive users of alcohol and those who take medicines which interfere with the magnesium metabolism, e.g. diuretics, heart medicine and antibiotics.

Various animal experiments were presented at the conference, one of which showed that magnesium supplementation was able to reduce high blood pressure. Another experiment, in which rabbits on a high cholesterol diet were given magnesium supplementation, showed that arteriosclerosis could be limited in these animals. Magnesium deficiency during pregnancy was reported to result in migraines and pregnancy-related high blood pressure, while a connection between magnesium deficiency and low birth weight, miscarriages and still births was also indicated.

Magnesium therapy
An imminent or existing magnesium deficiency caused by illnesses or medicines can easily be corrected with magnesium supplementation.

Furthermore it may well be worthwhile to try magnesium supplementation for a range of heart problems. This biological therapy is harmless and cheap and may alleviate symptoms in many cases. In addition, magnesium therapy may also reduce the damage in cases of heart infarction, which is why an increasing number of doctors are recommending magnesium as a routine component of therapy after a heart attack, partly because it diminishes the risk of fatal cardiac arrhythmia.

Several studies, in Denmark and Finland, amongst others, indicate that patients who suffer acute infarction of the heart have often had magnesium deficiencies over an extended period prior to the heart attack. Dr **B. B. Jeppesen**, from the cardiac department of Glostrup hospital in Denmark, delivered a report on this to the Scandinavian Trace Mineral Conference in Odense in 1987. A major study by the Finnish National Insurance Institution documented that magnesium deficiency in blood can often be detected as many as five years before heart infarction occurs. A Danish doctor, Dr **H. S. Rasmussen** has shown that magnesium therapy after a case of heart infarction helps to reduce the risk of a renewed attack.

Magnesium is considered to be a mildly tranquilizing substance, and, since it is perfectly harmless, it can replace tranquilizing medicines to a certain extent.

Undefined fatigue, and both the physical and psychological symptoms associated with it, can also be treated successfully with magnesium supplementation.

10.   IRON

Iron is more valuable than gold.

Finnish proverb

The commonest nutritionally induced deficiency disease in the industrialized world is anaemia (iron deficiency in the blood). The incidence of this disease would decrease if women would increase their intake of vitamin C, which improves the otherwise modest absorption of iron from foodstuffs. Iron improves general health and resistance, and protects us against infections and many other diseases.

Adults have about 3–4 g of iron in their bodies, and the greatest proportion of this (about 65%) is in the haemoglobin; 10% is concentrated in the myoglobin in the muscles, while the rest is stored in the liver, the spleen, the kidneys, the bone marrow and other organs. There are about 500 mg of iron in one litre of blood. Most of this quantity resides in the red blood corpuscles, although there is also a certain amount of iron in serum. Serum iron and transferrin values are normally used to ascertain the quantities of iron in the body.

Iron requirements and metabolism
The average intake of iron is too low, although many people receive more than 6 mg per 1000 kcal or 18 mg per day, which is the RDA. This dose covers the absolute

minimum requirement for an adult. It is generally accepted that, in many countries, iron intake should be considerably higher than it is, particularly amongst women.

Men and women over the age of 50 years are particularly susceptible to iron deficiencies.

Dietary iron has three main sources: the haem iron from meat, the non-haem iron from vegetables and the non-haem iron which is added to foodstuffs. The best dietary sources of iron are meat, offal, blood, peas, parsley and sea foods. 40% of the iron in meat is haem iron, which is easily absorbed, on average up to 24%. Other dietary components do not have much effect on the absorption of haem iron. Nevertheless, haem iron in the diet generally covers only 10% of the total intake of iron, the rest being poorly absorbed non-haem iron. This is found in unleavened bread, eggs and vegetables. A continental breakfast (toast, butter, marmalade and tea or coffee) yields only 0.15 mg absorbed non-haem iron. Corn flakes and milk do not contribute at all to the iron absorbed. A breakfast of egg and bacon increases the amount of absorbed iron to 0.25 mg, and a glass of orange juice, because of its vitamin C content, further increases the absorbed iron to 0.4 mg. These data were recently presented by Professor **Leif Hallberg** from Sweden.

Coffee and tea significantly inhibit the absorption of iron. The calcium in the milk competes with iron, the bio-availability of which is consequently lowered by drinking milk with the meal. Orange juice, on the other hand, is a superior drink which remarkably increases iron absorption from the meal. Professor Hallberg reminds us, however, that calcium is an equally important mineral, and he warns against totally abandoning milk because of its inhibition of iron absorption. It is possible to balance the absorption of both calcium and iron by intake of vitamin C-containing salad, juice or fruit and milk during meals.

Many foodstuffs are fortified with iron (non-haem). Since 1974 in Finland, for instance, iron has been added to wheat flour and breast milk substitutes. The added iron totals 17% of the average iron intake, and represents an additional 4 mg for adults and 2.5 mg for children, per day. The iron is added in the form of iron powder, in the reduced (ferrous) form. It is absorbed reasonably well and it does not react chemically with the other ingredients in the foodstuff. Presumably therefore, it does not cause lipid peroxidation in the food. In breast milk substitutes, soluble ferrosulphate is added, which is readily absorbed but which tends to react with the lipids in the milk leading easily to peroxidation. In Sweden, on the other hand, the added iron is in the form of ferric orthosulphate, which is more easily absorbed than reduced iron, at least from children's breakfast cereals, according to studies conducted by Professor Leif Hallberg.

The impending common market legislation in Europe from 1992 will force different countries to agree on levels and chemical forms of iron fortification in foodstuffs.

Anaemia is usually caused by unhealthy eating habits or an unvaried diet. Milk, for example, contains very little iron, and if a child drinks a lot of milk, it leaves less space for other foods with a higher iron content. Other reasons for anaemias are sparse intake of meat, banting (treatment involving sugar-, starch- and fat-free slimming diet), one-meal-a-day lifestyles and inadequate vitamin C intake. As a result, about one-third of the teenage girls and women in Europe have latent iron deficiency.

Heavy menstrual bleeding and haemorrhaging in the gastrointestinal tract (stomach ulcers, haemorrhoids, cancer) can result in significant losses of iron, which often leads to anaemia. Long-term infections and rheumatic complaints also tap iron reserves.

The requirement for dietary iron during pregnancy is 30–60 mg per day. Vitamin C supplementation doubles the absorption of iron and inhibits the development of anaemia. Very high doses of vitamin C should, however, be avoided towards the end of pregnancy because the foetus becomes accustomed to the high vitamin C content, and when the baby is later fed with breast milk, symptoms of vitamin C deficiency may arise.

Breast feeding also increases the mother's iron requirements. The iron content in the blood is usually measured at regular post-natal check-ups, and a need for iron supplementation is the rule rather than the exception. Children and adolescents not only need iron for the haemoglobin in their blood, but also for the building up of iron reserves in the body.

Iron absorption from food is not usually very efficient, mostly only about 5% is absorbed, except for those quantities which are absorbed from meat and offal. The cause of anaemia is therefore often not a lack of dietary iron, but rather a decreased absorption capacity. Only about 4% of the iron in some iron tablets is absorbed.

The absorbed iron arrives in the blood, where it binds itself to protein substances in the red blood corpuscles, the haemoglobin. For this process to take place it requires the presence of copper, cobalt, molybdenum and vitamin E. The absence of one of these factors may lead to anaemia, which can therefore not be corrected with iron supplements alone.

The haemoglobin transports oxygen from the lungs to all tissues, from where the cells utilize the oxygen as a source of energy. Iron is also transferred from the blood to the myoglobin in the muscles and other tissues which store iron. Myoglobin, like haemoglobin, is a protein which binds oxygen.

Iron loss is usually about 0.8–1.0 mg per day, the lower figures usually applying to women. During menstruation, however, up to 1.4 mg iron is lost and this can lead to anaemia in time. When we donate blood, a short-term, but harmless, anaemia occurs.

The effects of iron

The production of red blood corpuscles, oxygen transportation and the functioning of many enzymes in the organism require iron. In addition, the metabolism of B vitamins is dependent on iron. Iron is an oxidant, which oxidizes vitamins A, C and E and hastens the rancidification (lipid peroxidation) of fatty acids in foodstuffs and in the body.

Anaemia is one of the commonest diseases in the world. Symptoms of anaemia are tiredness, loss of strength, nervousness and pale skin. The skin is often itchy and the patient becomes rapidly out of breath. The haemoglobin content of the blood therefore has a direct influence on the capacity for physical activity. The latter falls in women when haemoglobin levels drop to less than $120–130 \text{ g l}^{-1}$; the pulse rate rises and lactic acid accumulates in the tissues.

Iron supplementation to a deficient person improves his or her capacity for physical performance so rapidly that it would appear that this is not only caused by an increase in haemoglobin levels. It is possible that iron has a number of properties which we still do not fully understand.

Iron deficiency also increases the risk of contracting infectious diseases, because the defence mechanisms of the white blood corpuscles (lymphocytes and granulocytes) deteriorate. In addition, the capacity of the T-lymphocytes to identify virus, bacteria and cancer cells is dependent on iron (see also 1.6).

A study in Chicago has shown that dietary supplementation of iron for children reduced the incidence of anaemia and infections of the respiratory passages. Similar results are reported from New Zealand, where Maori children were given iron injections.

The connections between iron deficiency, iron supplementation and infectious diseases should nevertheless be researched more thoroughly.

It can occur, although this is very rare, that an excess of iron accumulates in the tissues. One of the causes of this can be an excessive accumulation of red blood corpuscles, known as polycythaemia. Excessive amounts of iron also accumulate in the tissues with the rare diseases, haemosiderosis and haemochromatosis. Very high concentrations of iron in the organism can increase the receptivity to infectious diseases.

Iron therapy
Iron supplementation is usually necessary during pregnancy and in the event of iron-deficiency anaemia. Iron tablets can produce a number of side-effects, including constipation, diarrhoea, indisposition and stomach pains. Side-effects can usually be lessened if preparations of iron are taken after meals, but this may result in poorer absorption of the iron supplementation. If stomach problems arise, it is necessary to reduce the daily dose or to take it in smaller quantities several times a day. Iron tablets give the excreta a black colour.

Normally, in the event of anaemia, it can take up to six months to correct the iron balance in the organism. It can take longer, or may even be virtually impossible if there is severe menstrual bleeding or internal haemorrhaging due to a stomach ulcer, for example.

When the anaemia has been corrected, iron supplementation should be stopped, otherwise the risk of excessive reserves of iron in the tissues may arise.

Broad spectrum antibiotics, such as doxycycline, should not be taken together with iron supplementation, because these form insoluble compounds with minerals.

It is advisable to allow two hours to pass between the taking of iron supplementation and other mineral preparations. Iron competes with dietary calcium and manganese, for example, and they can therefore interfere with the absorption of iron. Conversely, long-term iron supplementation can lead to calcium and manganese deficiency.

11. ZINC

> *Preconceived opinions are not so foolish, as long as they are right.*
> Peter Hill (1854–1904)

Zinc is of great importance for the health of a number of population groups: growing

children, pregnant women, the elderly, and people with allergies and chronic diseases. They may actually benefit from zinc supplementation. Vegetarians are also one of the potential risk groups for zinc deficiency, and dyslexia is also associated with zinc deficiency.

Yet most doctors' knowledge about zinc leaves a lot to be desired, and it is only in the very recent past that the role of zinc has been acknowledged as a beneficial mineral for the treatment and prevention of diseases.

The adult human organism contains 2–4 g of zinc, of which the greater part (78%) is concentrated in the bones, the muscles and the skin. In the event of zinc deficiency, zinc is released from these tissues, so we are dependent upon a daily dietary supply of zinc. Zinc is an integral component of almost 100 different enzymes; it protects the skin and improves resistance to infectious diseases, inflammations and allergies.

Zinc requirements and metabolism

Opinions about zinc requirements are widely divergent. The adult RDA for zinc is 15 mg, but the actual intake of a great part of the population is considerably less, particularly among the elderly and among fast-growing adolescents on junk food. During pregnancy and breast feeding the requirements increase by 50–75%.

In Scandinavia and in many other parts of the world an average male diet of 2800 kcal provides less than the recommended 15 mg. If we eat less than the average calorie intake, our intake of zinc will be correspondingly lower.

The best dietary sources of zinc are meat, fish, liver, vegetables and oysters, although there are considerable variations in the bio-availability for zinc from different foodstuffs. It may be worthwhile to note that in a normal, mixed diet, about 15–40% of zinc intake is absorbed, while the zinc in vegetables is less readily absorbed. Vegetarians must therefore be aware of the possibility of developing zinc deficiencies, if they do not ensure a regular zinc supplement to their diet. The most readily absorbable form of zinc supplement is organic zinc, in the form of, for example, gluconate or aspartate.

Zinc is the antagonist of copper, with the result that a high zinc intake may reduce the absorption of copper. A zinc supplementation of more than 30 mg per day should be accompanied by a small supplement of copper.

Zinc is mostly eliminated from the organism with secretions from the pancreas and the gall bladder and is transported out of the body with the excreta. Zinc loss increases with a range of diseases and damage to the tissues, such as burns, kidney diseases, flaking skin diseases, diabetes, excessive sweat secretions, dialysis treatment, diuretic medicines and certain anti-arthritis medicines.

The effects of zinc

Zinc is vital to about 200 different enzymes, to the formation of bone tissue, in the healing of wounds and sores, to the production of proteins, the regulation of ribosomal, ribonucleic acid (RNA) synthesis and insulin and in the carbohydrate metabolism. Dietary zinc deficiency is known to inhibit the growth of children. Zinc is also an important participant in the copper–zinc superoxide dismutase enzyme (CuZnSOD), which has antioxidant properties. We will discuss this enzyme in more detail in the section on copper.

The most common symptoms of zinc deficiency are skin problems, because 20% of the zinc in the organism is concentrated in the skin, which reacts very quickly to deficiency conditions. Zinc deficiency also expresses itself in a loss of the sense of taste and in general fatigue. Other possible symptoms of zinc deficiency are hormonal disturbances and childlessness. Zinc deficiency may also associate with dysfunctions of the brain, in dyslexia for example (see below).

Zinc reinforces the immune defence from outside attack. The organism is constantly exposed to harmful viruses, bacteria, carcinogens and allergens. Zinc deficiency reduces the effectiveness of our resistance to these invading micro-organisms and substances.

Vitamin A metabolism is also dependent on zinc. The enzyme which transforms retinol to retinal requires zinc in order to be able to carry out this process. Zinc deficiency can therefore also reduce the organism's capacity to utilize supplies of vitamin A, leaving the immune defence lacking in two of its important components.

In addition, zinc is a prerequisite for the transformation of linoleic acid to γ-linolenic acid (GLA; see 6.2), which is, in turn, a precondition for the production of beneficial prostaglandins.

Zinc therapy

Zinc is of value in the treatment of skin wounds, because it accelerates the healing process. There are also indications that zinc could also be useful as adjunct therapy for pre- and post-operation patients, for the above-mentioned reasons. In addition, zinc is a component of many skin-protective ointments.

Zinc has also been administered in tablet form in the treatment of stomach ulcers, arthritis, acne and heart diseases, but without uniform success.

I have treated many patients with zinc and selenium, usually in combination with vitamins A, B6, C and E and as adjunct therapy together with other forms of treatment. The results of zinc therapy for acne, skin diseases and a range of allergic problems have been remarkably good. Many hay-fever patients have been able to dispense with antihistamines altogether, after zinc, selenium and vitamin E therapy.

Zinc is not particularly toxic, and serious side-effects are unlikely to arise, even with considerable overdosing with this mineral. However, when very high doses (over 30 mg per day) of zinc are used over lengthy periods, it is advisable to take an additional supplement of copper, in order to avoid possible cholesterol increase as well as copper deficiency.

Zinc may also be a valuable adjunct therapy for cancer patients, because it has an anti-metastatic effect (i.e. it helps to limit the spread of cancers). An appropriate dosage for this purpose would be 15–30 mg per day.

Zinc has attracted considerable attention in the medical world recently, because some studies have indicated that zinc deficiency may be connected with dyslexia. A British study, involving 26 schoolchildren between the ages of 6 and 14 years with reading difficulties, measured zinc values from sweat loss. The dyslexic children were discovered to have zinc values which were, on average, only 66% of those of their classmates without reading problems, who served as a control group.

This study, which was published in the *British Medical Journal*, considered the possibility that the source of the problem may lie in zinc deficiencies in one of the parents at the moment of conception. The findings of new studies on the significance of zinc supplementation for infants or even for the foetus were also emphasized.

The British study gave an impetus for my team to analyse blood zinc and selenium status in 18 dyslexic Finnish schoolchildren and to supplement them for one year with 15 mg of zinc and 100 μg of selenium per day. It turned out that 13 children out of 18 had lower than normal zinc concentrations in whole blood. Psychological tests showed that, after eight months of supplementation, 11 children in the group had improved significantly and had many fewer difficulties in reading and writing. According to the school teachers, a remarkable change for the better was observed, in the class in general, as early as after two months of supplementation.

The possible importance of zinc for the human nervous system has also been demonstrated in studies which have compared zinc levels in the placenta with the circumference of the baby's skull. Indications are that low zinc levels are associated with small skull circumference and corresponding delays in brain development.

Zinc has also been used experimentally in the treatment of the eye disease, senile degeneration of the macula, which is one of the commonest causes of impaired eyesight or even blindness in the elderly. This occurs because certain enzymes in the eye apparently lack zinc, so the eye's yellow spot fails to function properly, as a consequence of the harmful effects of free radicals. Experiments with zinc supplementation are being conducted at present.

12.   COPPER

There are two contradictions, which make us suspicious. The old one which does not work, and the new one with which we are unfamiliar.
                                                                            La Bruyère (1645–1696)

There are only about 80 milligrams of copper in the adult human organism, but it is nevertheless an immensely important cell-protective mineral. Most of the copper in our bodies is to be found in the liver, the muscles, the bones, the heart and the brain. The copper content in blood increases with diseases such as arthritis, infections and cancer. It would appear that copper may be of major significance in the treatment and prevention of precisely these diseases.

Copper, like iron, is as such an oxidant metal, and yet it has an antioxidant function in the organism in its role as a participant in the antioxidant enzyme SOD. This enzyme is now commercially available as a medicine. It can be administered in the form of an injection and has been experimentally applied for the treatment of inflamed joints.

Copper requirements and metabolism

The RDA for copper is 2 milligrams, but, in Finland, for instance, the average daily diet contains 1.7 mg of copper.

Dietary copper occurs in offal, shellfish, nuts, raisins and peas, and, unlike most other trace elements, it is absorbed in both the stomach and the small intestine. The copper balance is controlled by dietary intake and by variations in copper excretion in the faeces. If the copper reserves in the body are filled up, absorption can drop by 80–90%. Absorption capacity can be affected by a range of other dietary factors, such as cadmium, cobalt, molybdenum, iron, zinc, fibre and proteins. Copper absorption is reduced in particular by a high zinc intake, because the two minerals compete to form compounds with proteins. The presence of copper is, moreover, necessary for the binding of iron to haemoglobin.

In the blood, equal quantities of copper are found in red blood corpuscles and plasma. A proportion of the copper in the red blood corpuscles is loosely bound to amino acids. This part is called the labile pool or the metabolic storage. However, 60% of the copper in blood cells is tightly bound to the important enzyme copper–zinc superoxide dismutase (CuZnSOD), and this is known as the stable pool or the permanent storage. As yet, little is known about the copper metabolism or about how it passes from the stable to the labile pool.

CuZnSOD is an enzyme which contains two copper and two zinc atoms in immediate proximity to each other. This enzyme protects the cells from the damage caused by free radicals and peroxides. In order for the enzyme to be able to function, both copper and zinc must be present.

Almost all the copper in the plasma is bound to the protein, ceruloplasmin, which is manufactured in the liver. Ceruloplasmin functions as an antioxidant in the body. This is why the content of ceruloplasmin (and copper) in the blood increases with inflammations and other conditions where free radicals are formed in excess. The body tries to combat the dangerous lipid peroxidation by enhancing the production of endogenous antioxidants. Then copper content of plasma is usually 90–145 μg l^{-1} and in whole blood it is 85–125 μg l^{-1}. Women who use a copper contraceptive coil absorb 30–50 μg of this copper per day, or about 1% of the amount absorbed from the daily diet.

The blood levels of copper rise in connection with oestrogen treatment and during pregnancy. Copper values also increase with the inflammation reactions associated with, for instance, arthritis. Copper blood values can therefore be a reflection of the activity of this disease.

Copper is transported out of the body, together with bile, in the excreta. Under normal conditions, the amount of excreted copper is equal to intake.

The effects of copper
The copper–zinc enzyme

Copper participates in many enzyme systems, but it is best known for its active role in the SOD enzyme.

CuZnSOD is the fifth most common protein in the human organism. It is found in the red blood corpuscles and in all tissues, where it renders oxygen free radicals harmless (much in the same way as the selenium enzyme, glutathione peroxidase).

SOD enzyme has been tested recently in the treatment of inflamed joints, as occur in arthritis. In many cases a beneficial but short-lasting effect has been observed.

A comparative study in Marburg, West Germany, was able to show, for example, that SOD therapy was more effective than acetyl salicylates. With SOD the rheumatoid factor fell by 60% in the joint fluids, and only by 20% with acetyl salicylates. SOD also reduced the levels of PGE2, a harmful prostaglandin, more effectively than aspirin.

Positive effects have also been observed with SOD therapy for other types of inflammation of the joints, including osteoarthrosis. SOD would therefore appear to be a promising new natural medicine for the treatment of arthritis, although more controlled studies are required on this area.

Ceruloplasmin

Ceruloplasmin is a dark blue plasma protein which contains eight copper atoms. Ceruloplasmin regulates the levels of the hormones, adrenaline, noradrenaline, serotonin and melatonin in plasma. The levels of ceruloplasmin increase in the blood during physical activity.

This plasma protein is also required for the production of red blood corpuscles. Copper deficiency can lead to storing of iron in the liver, thus obstructing the transportation of iron to the blood cells. Furthermore, copper deficiency (and therefore of ceruloplasmin) can cause iron to be oxidized more easily and this may contribute to the formation of superoxide radicals in the cells.

Ceruloplasmin has an antioxidant effect and can also neutralize free radicals itself. If the content of ceruloplasmin in the blood falls, copper is stored in the tissues instead.

Cytochrome oxidase

This enzyme is extremely important for the metabolism, and its activity is reduced in the event of copper deficiency. The energy production in the cells falls and this can have far-reaching consequences for the entire food metabolism.

Copper therapy
Copper deficiency

Serious deficiencies of copper are rare, although they can arise in connection with serious kidney diseases, such as Wilson's disease, Menke's syndrome and various other disturbances in the absorption of food.

The risk of heart and circulatory problems would also appear to increase with copper deficiency, especially if this is accompanied by a deficiency of selenium. If the activity of both the important antioxidant enzymes, CuZnSOD and glutathione peroxidase, is impaired, this allows the free radicals to effect damage to the cells virtually unhindered. This danger is particularly imminent for people with constricted blood vessels, because ischaemia (oxygen deficiency) also promotes the production of free radicals (see section 2.4, on reperfusion injury).

Copper deficiency can furthermore contribute to anaemia, bone diseases, disturbances in the nervous system and hair loss. In children the bones may become brittle and growth is inhibited by copper deficiency.

Diseases of the joints and infections

The quantities of ceruloplasmin and copper in serum rise in the space of a few hours after the onset of an infection. Zinc values fall rapidly, simultaneously with a rise in copper levels in the event of tuberculosis of the lung or other infectious diseases.

Thus copper and zinc levels, like erythrocyte sedimentation rate (ESR) can be used as indirect parameters for hidden infections in any specific part of the body.

After the illness has subsided values for these minerals return to normal within a few weeks.

Cancer

Malignant tumours increase the blood levels of copper. This has been observed in cases of Hodgkin's disease, non-Hodgkin lymphomas and in cancers of the bladder, the lungs, the breast and the womb. On the other hand, no changes in copper values have been detected in cases of cancer of the prostate.

Increased blood levels of copper cannot, however, be assumed to be an unequivocal indication of cancer. Nevertheless it does represent grounds for further investigation.

The first effects of the medicinal treatment of cancer are normalized copper levels. Conversely, a rising copper content in the blood may be an indication of the onset of a relapse.

13.   MANGANESE

Never occupy yourself with fortune telling. If your predictions are incorrect, everyone will remember it, and if your predictions come true, everyone will forget it.

Josh Billings (1818–1885)

The amount of manganese in the adult human organism has been determined as 10–20 mg, which is mainly concentrated in the liver, the large intestine and the kidneys. Manganese is transferred, via the placenta, from the mother to the foetus, which has a manganese content of about half of the mother's. Manganese is an essential mineral and an antioxidant, which participates in the SOD enzyme system, like copper and zinc.

In some studies, deficiencies of manganese in Scandinavia have been connected with the incidence of cancer.

Manganese requirements and metabolism

The RDA for manganese is 3.8 mg, and a number of studies have indicated that actual daily intake is closer to 5–6 mg. There would therefore appear to be no question of general manganese deficiency.

Nuts, whole grains, vegetables and tea have a high content of manganese, while meat, fish and dairy products only contain insignificant quantities of this mineral. Only about 3% of the manganese in foodstuffs is absorbed in the organism. Manganese competes with iron, so that the absorption of manganese increases if there is a deficiency of iron.

Manganese is eliminated with the urine, and this loss is increased with alcohol. Manganese in the organism is mostly found within the cells and is stored in tissues which contain much pigment, i.e. dark skin, dark hair and the retina of the eye.

Manganese effects and therapy

Manganese is required in the formation of the bones, for the metabolism of fats and carbohydrates and for fertility. Dietary deficiency diseases associated with manganese are not known, although experimentally induced manganese deficiencies produce the following symptoms: emaciation, rashes and other skin changes and reduced hair growth. Manganese deficiency has been suspected as a contributory factor in diabetes and epilepsy, but these findings are unconfirmed by controlled studies.

There are no official recommendations for the use of manganese as therapy or in preventive medicine. Orthodox medicine does not accept that manganese deficiencies can arise under normal conditions.

Despite this, I have prescribed manganese on several occasions — after conducting blood tests — to patients with hampering pains in joints and bones, and this has had a beneficial effect in some cases. This has been accompanied by a normalization of blood values for this mineral. Manganese activates the organism's own killer cells (see 1.6, on the immune system) and I therefore prescribe manganese (800 mg manganese glyceryl phosphate per day) as adjunct therapy for my cancer patients.

14. MOLYBDENUM

Nobody can know everything.

Horatius (65–8 BC)

The significance of molybdenum for health is not very well known. Nevertheless, it is classified as an essential trace element because it is a component of at least three enzymes. These enzymes are active in the metabolism and neutralize toxic compounds of sulphur. Molybdenum is also of importance for the production of haemoglobin.

It has been discovered that molybdenum deficiency in ruminant (cud-chewing) animals inhibits growth and leads to loss of appetite and premature death.

Molybdenum requirements and metabolism

Human requirements for molybdenum are not known in any detail, although the RDAs in various countries range from 100 to 500 μg.

Good dietary sources of molybdenum are offal, grains and leguminous fruits, although molybdenum content in foodstuffs varies considerably, according to the natural content in soils.

In some places in the Soviet Union daily intake is as high as 10–15 mg, i.e. almost 100 times the normal daily intake. An arthritic disorder in the joints is common in these areas and this may be due to an excessive intake of this trace element.

Absorbed molybdenum is distributed throughout the different tissues in the organism, although the largest concentrations are in the liver, which appears to be able to store a certain quantity of this mineral. Molybdenum is eliminated with both urine and faeces but its metabolic processes are almost totally unknown.

The effects of molybdenum

There are no known diseases which can be directly connected with molybdenum deficiency. There are, however, areas in China where the molybdenum content of the soil is extremely low, and there is a high incidence of cancer of the oesophagus amongst the population in these areas. One theory has suggested that this is caused by the failure of the molybdenum enzyme, and that this leads to the accumulation of carcinogenic compounds of nitrogen in the tissues. Some animal experiments have supported this theory, but no real confirmation is available as yet.

Serum molybdenum levels are usually less than $1.1 \mu g \, l^{-1}$, although these values rise in the event of liver diseases, for reasons which are, as yet, unknown. A poor metabolism of this mineral can be observed in a number of hereditary diseases, where the enzymes, xanthine oxidase and sulphite oxidase do not function optimally. These diseases are very rare and molybdenum supplementation is unable to prevent them from resulting in morbidity.

It is generally believed that molybdenum has an effect in the prevention of dental decay, at least in the case of animals. Also, there are indications that the incidence of dental decay (caries) is lower in molybdenum-rich areas.

In contrast to the above findings, experiments with rats have come to the opposite conclusion: that molybdenum actually speeds up dental decay. This is the case with many of the research results on this mineral — the few available results are contradictory.

Molybdenum is an interesting trace mineral, but it is also one which needs to be researched more thoroughly. At the moment it does not figure in micronutrient therapy.

15. ARSENIC

The adult human organism has a natural content of about 14 mg of arsenic. Hitherto, this substance has been viewed exclusively as a poison, which healthy people were believed to have no need for. This view is now changing, and there is now some evidence which suggests that arsenic, in fact, may be an essential trace element. It has been discovered, for example, that experimental animals, whose dietary intake of arsenic is less than 0.5 mg per kg body weight, undergo disturbances in growth and fertility and that internal organs can be exposed to damage.

The human requirement for arsenic is not known exactly, but it is presumably between 0.5 and 4 mg per day. Arsenic is stored in all tissues, but particularly in hair and nails (it is from here that samples are taken in the event of a suspected poisoning).

Generally, we only really know about the toxic properties of arsenic, and no deficiency diseases associated with this mineral are known.

Arsenic is able to counteract the effects of possible selenium poisoning and vice versa. A recent study on Swedish coal miners has shown that a specific form of lung cancer, which is associated with environmental arsenic exposure, occurs more rarely when blood levels of selenium are high.

16. BORON

Boron may be of great importance in the prevention and treatment of osteoporosis and rheumatoid arthritis. Although further scientific and medical studies in this area are needed, there is a now a considerable body of evidence, much of it of good scientific

quality, which points to boron as an essential trace element. When boron is mentioned, most of us think of borax or the aqueous solution of boric acid which is used in the treatment of swellings and eye infections. Boric acid and borax have a feeble bacteriostatic action (i.e. they stop the growth of bacteria). It now seems that dietary boron may be one of the essential trace elements which human beings need. Boron is, in any case, essential to animals and plants.

There is no RDA for boron, but most people have a dietary boron intake of about 1–2 mg per day, although intakes as low as 0.25 mg have been reported. The mean dietary intake in Finland has been estimated at 1.7 mg per day. Boron is naturally present in all normal diets. There are abundant quantities of boron in plants and vegetables, but only very little in meat and other animal foodstuffs. Individual intakes therefore can vary widely, according to the proportions of the various foodstuffs in the diet. It has been estimated that 1–2 mg boron per day may meet human requirements.

A particularly good source of boron is wine. There is as much boron in a bottle of wine as there is in the human organism. In addition, apples, avocados, citrus fruits, pears, tomatoes, soya meal, prunes, raisins, nuts, dates, honey and berries are rich in boron.

Food group (Finnish)	Average boron contents (μg/g dry weight)
Cereals	0.92
Meat	0.16
Fish	0.36
Dairy products	1.10
Vegetable foods	13.00
Other foods	2.60

From Koivistoinen, P. (1980).

Boron levels in the food depend considerably on the source and hence the levels in blood and other tissues are sensitive to variations in the dietary intake. Due to the high boron content in vegetables, the vegetarian diet may be advantageous in preventing rheumatoid arthritis and osteoporosis.

Studies on the possible effects of boron on human beings have only been reported very recently. In Mauritius and Jamaica, low levels of available boron in the soil are associated with a high incidence of arthritis, whereas in Israel, with high boron soil levels, arthritis is uncommon. In New Zealand and Australia, boron, in the form of borax ($Na_2B_4O_7.10H_2O$), has been used in the treatment of rheumatoid arthritis and other diseases in the joints. Results suggest that there is a remarkable effect with this treatment.

Dr. **R. E. Newnham** in New Zealand has reported that a daily supplement of 6–9 mg of boron may reverse the symptoms of 80–90% of patients within a few weeks, with no side-effects. Significantly lower boron levels in the bones of rheumatoid arthritis patients compared to healthy controls support these findings. Dr Newnham also reported that boron can completely cure arthritis in horses, cattle and dogs. In New Zealand several people take boron supplements and are convinced of the beneficial effects. The anecdotal evidence should give an impetus to properly controlled clinical trials.

A study at the US Department of Agriculture by Dr **F. H. Nielsen** and colleagues showed that in post-menopausal women a daily supplement of 3 mg boron considerably reduces the excretion of calcium, magnesium and phosphorus in the urine. Furthermore, within eight days, boron supplementation (3 mg/day) increases oestrogen levels in blood by 100%. Oestrogen elevation is of great importance for menopausal women as this hormone is known to reduce demineralization of bones after the menopause.

These studies point strongly to the possibility that dietary boron deficiency may play a role in osteoporosis in post-menopausal women. The already available evidence suggests that boron supplementation (3 mg/day) would be advantageous for both the prophylaxis and treatment of osteoporosis and rheumatoid arthritis. It further appears that boron therapy may also be of benefit in the treatment of allergic complaints. It has been discovered that people with allergic rashes have low boron values in the skin.

17. BROMINE

In the past, bromine has been used in place of tranquillizers, and from the early 1950s it was used as anti-epilepsy medicine. It has since been discovered that bromine can cause brain damage and it is no longer used in modern medicine.

There is a small and harmless bromine content in diets of about 1.5–2.5 mg.

18. IODINE

Most of the iodine in the human organism (20–50 μg) is stored in the thyroid gland. It was discovered as far back as 1818 that iodine plays a role in the regulation of the activity of the thyroid gland, but another 100 years passed before iodine was used for the treatment of underdeveloped children with thyroid gland failure. In Finland, iodine deficiency was widespread in the past, but since the addition of iodine to table salt began about 50 years ago, this problem has more or less been eradicated.

The RDA for iodine is 100–200 μg per day. Most people do not require more iodine than the amounts which can be taken in with table salt and other foodstuffs.

A third of absorbed iodine is immediately transported to the thyroid gland while the rest is eliminated with the urine, possibly after having had effects on the organism which are still unknown.

Iodine tablets can be used to reduce the dangers of radioactive radiation for people who have been exposed to excessive radiation (like after the Chernobyl disaster), although under normal circumstances there are seldom grounds for preventive therapy with this mineral.

In the thyroid gland, iodine is bound to tyrosine, which is a precursor to the thyroid hormones, T3 and T4, which regulate the metabolic processes in the organism.

19. CHLORINE

Chlorine is an essential trace element. Our intake of chlorine is usually too high rather than too low because of our over-consumption of table salt. The adult human organism usually contains about 75 g of chlorine, of which about 60 g is found outside of the cells in blood plasma and tissue fluids.

Chlorine is closely connected to sodium, and regulates the fluid balance, the acid–base balance, as well as directing the osmotic pressure. (Table salt is sodium chloride, NaCl.)

Chlorine is a component of stomach acids (hydrochloric acid, HCl) and therefore an important factor in the digestion. The carbon dioxide transport in the blood is also dependent upon chlorine.

20. SILICON

It is better to know a little about everything than everything about one thing.
                                                        Blaise Pascal (1623–1662)

Silicon is a 'new' essential trace element for both human beings and animals. The presence of silicon is a precondition for normal growth and development and for the formation of the bones. The lower our calcium intake is, the more important it becomes that we receive adequate quantities of silicon. No actual deficiency disease is known of in connection with silicon deficiency, but a diet which is rich in silicon strengthens the nails, the hair, the bones and other connective tissues. The main dietary sources of silicon are milk and vegetables.

In nature most of the silicon occurs as silicon dioxide, which dissolves in water and forms silicates. Our silicon requirements are still unknown and there are no official recommendations for silicon intake. The organism contains about 7 g of silicon, far more than our copper or iron content and the daily silicon requirement probably lies in the region of about 20–30 mg. In Finland the average silicon intake is 29 mg per 2800 kcal. It is difficult to determine whether or not we are exposed to a risk of silicon deficiency because of the difficulties involved in measuring blood values of this mineral and because optimal levels of silicon intake are still unknown. No actual deficiency disease has been connected with silicon deficiency, but many of our modern refined foods are very low in silicon.

Silicon compounds are absorbed from the intestines, although we still do not know in which chemical form this takes place.

Inorganic silicon is not nearly as effective in raising blood levels for this mineral as organic plant silicon compounds. Silicon is transported with the blood to the bones, cartilage, muscles, sinews, nails and the brain. The lymph glands also contain a great deal of silicon. Most of our silicon intake is eliminated with the urine.

The effects of silicon

Silicon has attracted particular interest because of its capacity to stimulate bone-forming cells and for its role in the formation of the basic material of the connective tissues. This is why silicon has become popular in recent years for the strengthening of bone, hair and nail growth. Silicon is found in high concentrations in the bone-building cells, the osteoblasts, and plays an important role in the activity of these cells. Silicon is also required in the formation of the connective tissue proteins, collagen and elastin, which give connective tissues their respective contractive and elastic properties.

The effects of silicon on the hair and nails have not yet been documented in controlled studies, whereas it has been generally accepted that this mineral may have a beneficial effect on bone formation. Silicon (plus fluorine and boron) supplementation may therefore be recommended in the treatment of slow healing after a fracture. Similarly, people who have been fitted with an artificial hip after a fracture or because of osteoarthrosis, may benefit from silicon and boron therapy, which aids the growth of the new bone mass while it grows round the new joint and holds it in position. This is of particular value for the elderly, whose bones do not grow as fast as those of young people.

The bone-forming property of silicon raises the exciting question of whether or not silicon may be of use in the prevention or treatment of osteoporosis. This topic has not yet been adequately researched and, at the moment, any recommendations to this effect are based upon the theoretical knowledge about the different ways in which calcium, boron, magnesium and silicon contribute to the formation of the bone mass. It is not inconceivable that these minerals complement each other in the process of bone formation.

There would appear to be a connection between silicon deficiency and arteriosclerosis, because the silicon values in the artery walls drop with age and the onset of arteriosclerosis. It is not known, however, whether this reduction in silicon values is a cause or an effect of hardening of the arteries.

Recent studies in the field of veterinary medicine indicate that silicon may be effective against inflammatory reactions. My own experiences suggest that silicon supplementation may be beneficial in the treatment of inflammatory skin disorders, such as atopic eczema, and in the strengthening of the hair, nails and skin.

21. NICKEL

Nickel is an essential trace element, but although several deficiency diseases have been found in animals, none have been detected in human beings in connection with this mineral. Nickel is apparently involved in the protection of the cell membranes, but can also cause contact allergies and cancer.

Nickel requirements and metabolism

Our intake of nickel is entirely dependent upon our diet, and since this mineral occurs mostly in vegetables, vegetarians tend to have a significantly higher nickel intake than people on a mixed diet.

Although nickel is classified as an essential trace element, actual requirements are not known. In Scandinavia daily intakes of nickel vary from 130 to 400 μg, with the lowest figures applying to Finland. Whether the lower Finnish intake is of any significance is not known.

Nickel absorption from the intestines into the blood stream is very poor. Only about 10% is absorbed and this is not affected by increased intake. Nickel is mostly transported out of the body with the faeces, although small quantities are lost with the urine and some is also lost during heavy sweating. Serum nickel values are $1.5–5 \, \mu g \, l^{-1}$.

The effects of nickel

In experimental animals, nickel deficiency leads to a range of illnesses. Hens undergo changes to the liver, the skin and the blood, while rats and pigs experience reduced growth and decreased fertility. No actual deficiency diseases are known in human beings, but it has been discovered that nickel levels in the blood increase when the heart suffers oxygen deficiency (ischaemia). The origin of this phenomenon is still unknown, because it has been established that the released nickel does not come from the heart muscle itself. Attempts have been made, however, to utilize these variations in blood nickel values in the diagnosis of ischaemic heart disease and to ascertain the degree of damage after a heart attack.

At the moment, there is no known reason for the use of nickel in the prevention or treatment of illness. On the contrary, there may even be grounds for avoiding too much nickel, because nickel is known to cause allergic eczema.

A new Danish study has pointed to a possible connection between high blood nickel values and arthritis. In this connection it is worth mentioning that large quantities of nickel are released into water from metal pipes. The nickel content of the first water from the tap in the morning can be extremely high — it is therefore advisable to let the water run for a while first.

Nickel in the form of nickel carbonyl compounds in the working environment can be carcinogenic.

22. ALUMINIUM

There is currently considerable interest in the possible health risks arising from human exposure to aluminium. This is largely due to the suggestion that it may be a major factor in Alzheimer's disease, the predominant cause of senile dementia with up to 500 000 patients in the UK.

Aluminium requirements for human beings are not known and it is debatable whether it is of any benefit to humans. The average daily Finnish diet contains 7 mg of aluminium. Total daily intake of this mineral is 80 mg in Norway and more than 100 mg in the United States. The latter two figures include the amounts of aluminium which come from the surfaces of cooking pots, other kitchen utensils and metal-coated disposable juice cans. We are also exposed to airborne aluminium as a result of environmental pollution. These quantities of aluminium are not insignificant.

Acid rain lowers the pH of the soil and then aluminium is leached from the soil into the groundwater. The soil is abundant in aluminium; in fact, at 8% aluminium is the third most frequently occurring element in the Earth's crust.

In the UK and Norway it was discovered recently that concentrations of aluminium in drinking water correlated with the incidence of Alzheimer's disease.

Tea bags contain a lot of aluminium, 128 mg per 100 g (dry bag material). Other sources of aluminium include instant coffee, dried milk and table salt. Aluminium is also found in food additives, e.g. E173, aluminium 'silver' finish used in the manufacture of pills, such as buffered aspirin. Anti-acid tablets also contain high concentrations of aluminium. Ingestion of such a tablet causes an 800% increase in the urinary excretion of aluminium and a 125% increase in zinc excretion. Several vaccines contain aluminium.

Effects of aluminium

Aluminium is chemically extremely reactive and it has a tendency to bind other substances. Chemically, aluminium (Al^{3+}) resembles iron (Fe^{3+}). Aluminium is therefore able to change places with Fe^{3+} in iron-binding proteins, especially ferritin, and hence aluminium is able to enter the cells. Aluminium causes cross-linkages of collagen, it alters the calcium metabolism and it causes demineralization of the bones. Aluminium causes zinc loss, which may contribute to the development of dementia. If you are 50 years of age, your chances of contracting Alzheimer's disease are 1 in 200. When you are 80 years old, the risk is 1 in 20.

In dyslexia, high aluminium and low zinc levels have been discovered. In zinc deficiency, aluminium is able to bind zinc-carrying proteins; when zinc deficiency is corrected, the proteins will be saturated with zinc and there will be no place for aluminium to enter.

Aluminium hydroxide, which is an ingredient of stomach medicine, binds and inhibits the absorption of phosphates. Patients who are fed with a drip or who undergo dialysis treatment have a much higher absorption of aluminium, and this can give side-effects of bone and brain damage. Aluminium poisoning has even resulted in fatalities, which is why dialysis fluid is usually purified of aluminium.

23. SILVER

He who has no silver in his purse should have it in his tongue

Thomas Fuller (1608–1661)

Silver is only used externally as a medicine against warts. Silver is not required in metabolic processes and there is no RDA for this mineral. Experimental animals only absorb about 10% of orally administered silver. Absorbed silver is transferred to the liver and stored there. Whether or not this also applies to human beings is not known.

Silver is corrosive and has reciprocal effects with selenium, copper and vitamin E. In principle, it is thus able to protect the organism from overdoses of selenium. Conversely, selenium, copper and vitamin E may protect against the toxic effects of silver exposure.

People with chronic silver poisoning have a permanent blue colouring of the connective membranes in the eye. Compounds of silver are applied externally for the treatment of warts, to stop nose-bleeds and as a disinfectant.

24. GOLD

Gold is the king of kings.

Antoine de Rivarol (1753–1801)

Like silver, gold is not an essential mineral for human beings. The Frenchman, Jacques Forestier, began to treat arthritis patients with gold about 55 years ago. He believed that rheumatoid arthritis was a form of tuberculosis and that such a serious illness would have to be treated with the most expensive medicine possible. Today, gold has indeed a central role in the treatment of arthritis. There are tablets containing gold which give fewer side-effects than the dreaded gold injections.

Gold is a fine example of a trace element which is effective in the treatment of serious illness, although this may seem strange, because it is totally foreign to the organism and plays no role in the metabolism.

Healthy people have no physiological need of gold. Salts of gold are very poorly absorbed in the gastrointestinal tract. Today it is possible to manufacture chelate preparations of gold, which can be taken orally and which are easily absorbed in blood and tissues.

Intramuscular injections of gold give high gold concentrations in the tissues. 5% of the injected gold is excreted in the course of one day, while after a week 85% of the administered dose still remains in the body. Gold can be observed in the urine as long as one year after treatment.

Gold is used medicinally primarily in the treatment of arthritis because it seems to dampen down inflammatory responses in the joints. The mechanism of the therapeutic effect is not fully understood, but, according to a Japanese study, gold functions as an *antioxidant.* Supplementation with other mineral antioxidants such as selenium and zinc for arthritis patients is in no way in conflict with gold therapy.

Because gold is alien to the organism, it gives some patients side-effects in the form of itching, rashes and serious kidney damage. Kidney damage is not necessarily connected with the levels of gold concentrations in the organism, but rather with the degree to which the patient is sensitive to the treatment.

Metallic gold has previously been used by dentists for filling teeth, and has also been employed in surgery and in the treatment of skin diseases and asthma, but it is no longer used in these contexts.

25. TIN

The road to civilization is paved with tin cans.

Elbert Hubbard (1856–1915)

Tin is one of the essential trace elements. Rats need 1–2 mg tin per kg bodyweight for normal growth and development. Lower intake retards growth and leads to falling hair, eczema and reduced muscle power in the space of a few weeks. Tin is a component of many organic and inorganic compounds with various properties.

Tin requirements and metabolism

The human requirement for tin has not yet been quantified and there is therefore no RDA for this mineral. Average intake with food is about 1 mg of tin per kg of food, and this quantity is doubled in connection with deep-frozen and tinned food. Our tin intake is therefore greatly dependent on the extent of tinned food consumption, varying from 0.2 to 17 mg per day. Tin intake is probably highest in the USA where tinned food constitutes about 15% of total food intake. Food which is kept in open tins has an extra high tin content, because tin is released on being oxidized in the air.

Inorganic salts of tin are very poorly absorbed in the digestive system, compared with the absorption of organic tin. Organic and inorganic tin are also eliminated differently: organic tin is mostly lost with the faeces and inorganic tin with the urine.

The effects of tin

Tin participates in the absorption of a number of other minerals and trace elements. Thus it inhibits the absorption of copper and zinc in some tissues and can also limit the excretion of these minerals. On the other hand, it increases the elimination of zinc with the urine. It would, however, not appear to have any influence on the elimination of iron, manganese or magnesium.

Tin seems to be of significance in the metabolism of bone tissue, possibly through some effect of vitamin D.

Animal experiments have indicated that tin compounds may have a range of immunological effects: some of which strengthen and some of which weaken the functioning of the immune system. The growth of cancerous tissue may be inhibited to a certain extent by tin supplementation. It is not known whether this effect is mediated by the immune system or whether tin affects the cells directly.

Tin is scarcely toxic in itself, but there is still not enough evidence to support the medical use of tin in various treatments. The immune reinforcing and cancer-preventive properties of tin may become relevant in the future.

26. VANADIUM

Vanadium is one of the essential trace elements which we know least about. Vanadium is found in most tissues but in as yet unknown quantities. Nor is the extent to which vanadium occurs in diets known. Some plants, such as radishes and dill, contain a great deal of this trace element.

Vanadium deficiency has been discovered to lead to retarded development in chickens, and, surprisingly, to decreased cholesterol levels.

The significance of vanadium for human health is still unknown, but it may be that it can play a role in the treatment of cancer.

Vanadium requirements have not yet been established, but it is estimated that they are in the region of 1–4 mg per day. Only 1% of dietary vanadium is absorbed, the greater part of which is then eliminated with the urine. What actually happens with the vanadium which is absorbed is unknown, but serum vanadium values are less than $10 \mu g \, l^{-1}$.

In 1971, **Klaus Schwartz** and co-workers showed that rats which had suffered from long-term deficiencies of vanadium gained around 40% in weight when they were given vanadium supplementation, in the form of sodium orthovanadate.

Experimental animals develop symptoms from vanadium deficiency, although these observations are still unconfirmed and cannot be transposed to human beings.

Interim results indicate that vanadium can inhibit the occurrence of spontaneous tumours in animals and, in addition, may reduce the carcinogenic properties of certain chemicals. In some studies, vanadium salts have also been shown to be capable of destroying cancer cells. None of these results have been proved with any degree of certainty.

27. TUNGSTEN

Tungsten usually occurs in nature in compounds of iron, which are used industrially in the production of steel and dyes. It has not been established whether or not tungsten is an essential trace element for animals or human beings. There is no available information about average daily intake, although tungsten is a dietary component in mixed diets. Tungsten is stored mostly in the bones, the liver and the kidneys.

The significance of tungsten in the organism is associated with its reciprocal relationship with other trace elements, primarily molybdenum and copper. It can inhibit the absorption of molybdenum from the intestines and affects the activity of the molybdenum-dependent enzyme, xanthine oxidase.

Tungsten is hardly toxic, although industrial tungsten carbide has been revealed to cause lung damage.

There are no established norms for tungsten therapy in treatment or prevention of disease.

28. COBALT

Cobalt is a component of vitamin B12 and the adult human organism contains about 1 mg of this trace element. Good dietary sources of cobalt are buckwheat, figs and green vegetables. Its function in the organism resembles that of iron, but which essential functions it participates in are not yet fully known, apart from its role in vitamin B12. About 40% of the cobalt in the organism is found in the muscles and 14% in the bones. Cobalt deficiencies in human beings are not known.

There is no RDA for cobalt, partly because we are generally believed to have a sufficient intake and partly because the effect of cobalt in the organism is virtually unknown.

Attempts have been made to use cobalt in the treatment of various forms of anaemia in the hope of improving the activity of the red blood corpuscles. Similarly, cobalt has been tried as a means to reduce high blood pressure but this treatment has never become generally accepted.

Human beings can tolerate large quantities of cobalt, but if the daily intake exceeds 20–30 mg per day there is a risk of side-effects, involving thyroid failure and weakening of cardiac functions.

LITERATURE

Clausen, J. and Nielsen, S. A. Comparison of whole blood selenium values and
    erythrocyte glutathione peroxidase activities of normal individuals on supple-

mentation with selenate, selenite, L-seleno-methionine, and high selenium yeast. *Biol. trace elem. res.* **15** (1988) 125–138.

Clausen, J., Egeskov Jensen, G. and Nielsen, S. A. Selenium in chronic neurological diseases: Multiple sclerosis and Batten's disease. *Biol. trace elem. res.* **15** (1988) 179–203.

Davies, S. and Stewart, A. *Nutritional Medicine.* Pan Books Ltd, 1987, p. 543.

Dworkin, B. *et al.* Abnormalities of blood selenium and glutathione peroxidase activity in patients with Acquired Immuno-deficiency Syndrome and AIDS-related complex. *Biol. trace elem. res.* **15** (1988) 167–177.

Fabris, N., Amadio, L., Licastro, F. *et al.* Thymic hormone deficiency in normal ageing and Down's syndrome: is there a primary failure of the thymus? *Lancet* **1** (1984) 983–986.

Frolkis, V. V., Frolkis, A. A., Dubur, G. Y., Khemelevsky, Y. V., Shevchuk, V. G., Golovschenko, S. F., Mkhitarjan. L. S., Voronkov. G. S., Tsyomik, V. A., Lysenko, I. V. and Poberezkina, N. B. Antioxidants as antiarrhythmic drugs. *Cardiology* **74** (1987) 124–132.

Gey, K. F. On the antioxidant hypothesis with regard to arteriosclerosis. *Bibl. Nutr. Dieta.* **37** (1986) 53–91.

Hamrin, K., Jameson, S., Hellsing, K. and Lindmark, B. Survival of acute cerebro-vascular disease as related to functional activity index and to zinc, copper, iron, hemoglobin and albumin levels on admission. *J. Trace Elem. Exper. Med.* **1** (1988) 23–31.

Hayaishi, O., Niki, E., Kondo, M. and Yoshiwaka, T. (eds). *Medical, biochemical and chemical aspects of free radicals.* Vols 1 & 2. Elsevier, Amsterdam — New York — Oxford — Tokyo, 1989, p. 1559.

Kaplan, J., Hess, J. H. and Prasad, A. S. Impaired Interleukin-2 production in the elderly: Association with mild zinc deficiency. *Ibid* 3–8.

Koivistoinen, P. (ed.) Mineral composition of Finnish foods. Acta agricult. Scand. Supp. 22 (1980) 171 p.

Levander, O. A. Progress in establishing human nutritional requirements and dietary recommendations for selenium. Selenium Congress, Tübingen, July 1988.

Luoma, P. V., Sotaniemi, E. A., Korpela, H. *et al.* Serum selenium, glutathione peroxidase activity and high density lipoprotein cholesterol — effect of selenium supplementation. *Res. Commun. Pathol. Pharmacol.* **46** No. 3 (1984) 469–472.

Macarthy, M. S., Rationales for micronutrient supplementation in diabetes. *Med Hypotheses* **13** (1984) 139–151.

Marniemi, J., Niskanen, J., Maatela, J., Mäki, A., Seppänen, A. and Aromaa, A. Mineral and trace element levels in serum of men died of different diseases. Trace Elements in Human Health and Disease. Second Nordic Symposium, 17–21 August 1987, Odense University, Odense, Denmark.

Prasad, A., Clinical manifestations of zinc deficiency. *Ann. Rev. Nutr.* **5** (1985) 341–363.

Ringstad, J. and Fønnebø, K. The Tromsø heart study: serum selenium in a low risk population for cardiovascular disease and cancer and matched controls. *Ann. Clin. Res.* **19** (1987) 351–354.

Salonen, J. T., Salonen, R., Seppänen, K. *et al.* Relationship to serum selenium and antioxidants to plasma lipoproteins, platelet aggregability and prevalent ischaemic heart disease in Eastern Finnish men. *Arteriosclerosis* **70** (1988) 155–160.

Santavuori, P., Heiskala, H., Westermarck, T. *et al.* Experience over 17 years with antioxidant treatment in Spielmeyer-Sjögren Disease. *Am. J. Med. Genetics.* Suppl. 5 (1988) 265–274.

Schrauzer, G. N. (ed.). *Selenium — Present status and perspectives in biology and medicine.* Humana Press, Clifton, N.J., 1988.

Speich, M. Plasma zinc and myocardial infarction: New hypothesis. *J. Am. Coll. Nutr.* **6** No. 2 (1987) 187–188.

TEMA-6, Asilomar, 31 May — 5 June, 1987. Abstracts.

Tolonen, M., Halme, M. and Sarna, S. Vitamin E and selenium supplementation in geriatric patients. A double-blind preliminary clinical trial. *Biol. Trace Elem. Res.* **7** (1985) 161–168.

Tolonen, M., Sarna, S., Halme, M. *et al.* Antioxidant supplementation decreases TBA reactants in serum of elderly. *Biol. Trace Elem. Res.* **17** (1988) 221–228.

Trace elements in human health and disease. Second Nordic Symposium, 17–21 August 1987, Odense, Denmark.

Varo, P. *et al.* Selenium intake and serum selenium in Finland: effects of soil fertilization with selenium. Selenium Congress, Tübingen, July, 1988.

Wallach, S. Clinical and biochemical aspects of chromium deficiency. *J. Am. Coll. Nutr.* **4** (1985) 107–120.

6

Essential fatty acids

1. WHAT ARE ESSENTIAL FATTY ACIDS?

Truth is the child of time.

John Ford (1586–1640)

In the past, the essential fatty acids have been known as F vitamins. These substances are precursors to the prostaglandins, which are highly relevant to the debate on dietary fat content. Recently, it has been discovered that these polyunsaturated fats incorporate a range of new and very interesting effects on human health. They help, for example, to reduce high blood cholesterol levels and decrease the risk of blood clots. γ-linolenic acid and fish oils can help patients with atopic eczema, asthma, high blood pressure, arthritis, psoriasis, menstrual problems, disseminated sclerosis and cancer.

The clinical use of fatty acids is gradually becoming more widespread in the prevention and treatment of illness — as is the case with vitamins and minerals. Essential fatty acids are naturally occurring substances and therefore can be combined well with vitamin and mineral therapy.

Essential fatty acids resemble minerals in two ways:

(1) *They are necessary for the cells and for good health.*
(2) *The body is unable to produce them itself and they must therefore be consumed via diet (or supplements).*

A curious aspect of the debate on vitamins and minerals has been that many doctors, representing orthodox medicine, and medical authorities maintain that the RDA's for vitamins and minerals are correct (see 1.7). These recommendations are mainly based on animal experiments.

The opposite is the case for essential fatty acids. Orthodox medical authorities recommend that at least 10–15% of our intake of calories should come from essential fatty acids, despite the fact that animals can suffice with 1% is sufficient.

The first essential fatty acid was discovered in 1929, but until a few years ago only three were known. There are now a dozen known essential fatty acids; and linoleic acid and linolenic acid can be transformed into all of them through a series of metabolic processes. The best dietary sources of fatty acids are:

(1) *Linoleic acid*   Seed oils from borage, blackcurrants, evening primrose, sunflowers, maize and peanuts.
(2) *γ-linolenic acid (GLA)*   Breast milk, borage seeds, blackcurrant seeds and evening primrose seeds.
(3) *dihomo-γ-linolenic acid (DHGLA)*   Breast milk and offal.
(4) *arachidonic acid (AA)*   Meat, dairy products, marine algae and prawns.
(5) *α-linolenic acid*   Vegetables, soya beans and linseeds.
(6) *eicosapentaenoic acid (EPA)*   Fatty fish (herring, mackerel, salmon), fish liver oil and marine algae.
(7) *docosahexanoic acid (DHA)*   Fish, fish liver oil and marine algae.

In the medical literature these fatty acids are included in the abbreviated term, EFA (essential fatty acids). It is only in very recent years that research on these substances has begun in increasing earnest and today they are being used in the prevention and treatment of a range of illnesses. At the moment the prime interest is centred on the possible connection between EFA and heart disease, disorders of the central nervous system, allergies and failing immune functions.

What is it that makes essential fatty acids so important? First and foremost is the fact that all cell membranes in all tissues consist of fatty acids. These membranes are vital to cell functions and a deficiency of fatty acids can lead to serious disorders in the tissues. Another point, which confirms the importance of these substances for the human organism, is that essential polyunsaturated fatty acids are precursors of prostaglandins (PG) and leukotrienes. The metabolisms of these hormone-like substances are of great importance in the treatment and prevention of many diseases. For example, the prostaglandins regulate the calcium transport in the body, the content of high energy substances (cyclical nucleotides) and the activity of key enzymes in the cells.

Some prostaglandins (PGE1 and PGE3) and leukotrienes are beneficial to health, while others are harmful. PGE1 inhibits the aggregation of blood platelets and helps to prevent blood clots in this way. It also helps to expand blood vessels and prevent infections. In addition to the above, prostaglandins are also involved in the regulation of the activity of the T-lymphocytes in the immune system (see 1.6).

Arachidonic acid (AA) can become both a beneficial and a harmful prostaglandin. The harmful form can give rise to inflammation reactions in the tissues, but it can also be transformed into prostacyclin (PGI2), which, like PGE1, helps to prevent blood clots and expand the blood vessels. There is one more harmful essential fatty acid, thromboxane A2 (TXA2), which can cause contraction of the blood vessels, thrombocyte aggregation and a deterioration in inflammation conditions.

In order to be able to prevent and treat various illnesses, it is vital that we are able to regulate the metabolism of essential fatty acids, prostaglandins and leukotrienes. This can be achieved fairly easily with acetylsalicylates, modern anti-inflammatory medicines (steroids and non-steroids) but also with vitamins, minerals and EFAs.

For example, vitamin E can inhibit the formation of harmful prostaglandins, while selenium, zinc, and vitamins B and C are necessary for the production of beneficial prostaglandins.

2. γ-LINOLENIC ACID (GLA)

The truth is subjective.

Søren Kirkegaard (1813–1855)

The most important dietary essential fatty acid is linoleic acid, which occurs in large quantities in plant oils. However, linoleic acid is biologically inactive and cannot be used by the organism in its naturally occurring form. Linoleic acid must first be catalyzed into γ-linolenic acid (GLA) under the influence of the enzyme, delta-6-desaturase (d6d). A number of circumstances can disturb the effective functioning of this enzyme, including certain foodstuffs, stimulants, medicines and diseases, and this can prevent the transformation of linoleic acid to GLA.

A diet rich in plant oils does not necessarily protect against deficiencies of GLA, which should therefore be taken in the form of a food supplement. This essential fatty acid has been successfully used in the treatment of allergies, arthritis and related disorders, skin diseases and many other ailments which are associated with imbalances in fat metabolism.

The transformation of linoleic to γ-linolenic acid can be affected by so many factors that it seems not unlikely that most people have an insufficient intake of this essential fatty acid. The most important of these inhibitive factors are:

1. High dietary content of saturated fats.
2. Cholesterol.
3. Polyunsaturated fats.
4. Insulin deficiency (diabetes).
5. Alcohol consumption.
6. Viral infections.
7. Cancer, radiation- and chemotherapy.
8. Deficiencies of zinc, magnesium and vitamin B6, and possibly other vitamin and mineral deficiencies.
9. Smoking.
10. Ageing.

In addition, a range of medicines inhibit the production of GLA from linoleic acid, including acetylsalicylates, anti-arthritis medicines, cortisone and beta-blocking agents. Undernourishment and nutritional imbalances, such as excess protein and sugar consumption, also contribute to reduced GLA activity, which diminishes naturally after the age of about 50 years.

If the organism is unable to produce enough GLA itself, then it can be taken in the form of a food supplement. GLA is extracted from plants which are rich in this fatty acid. The highest natural content is found in the medicinal herb, borage (*Borago officinalis*), 25% of which is composed of GLA. Blackcurrant seed oil contains rather smaller quantities (14%) of this essential fatty acid, while evening primrose contains 7–9% GLA.

However, in order to gain the optimal benefit from GLA, it is usually not enough just to take capsules of the substance, without the presence of other substances which are essential to the functioning of the GLA enzyme, i.e. zinc, selenium and vitamin B6. The organism's requirement for GLA can therefore scarcely be met by a balanced, mixed diet alone, because, apart from the few plants mentioned above, GLA only occurs naturally in breast milk.

When the organism's production of GLA is facilitated, however, the production of the beneficial prostaglandin PGE1 increases. This prostaglandin has a number of beneficial effects:

- It reduces the tendency of blood platelets to aggregate and thereby reduces the risk of blood clots.
- It expands contracted blood vessels, which may alleviate pains associated with angina pectoris.
- It expands the respiratory passages, prevents mucous formation, infections and asthma attacks.
- It reduces cholesterol production.
- It reinforces the effects of insulin.
- It improves the activity of the immune system (primarily via its influence on the T-lymphocytes).

The improved functioning of the T-lymphocytes can be of great importance in the event of disturbed immune functions with, for example, ulcerative colitis, arthritic and related disorders, allergies, asthma and skin diseases.

GLA therapy can apparently also help with the following problems:

- Disorders in the central nervous system; in schizophrenia and disseminated sclerosis the patient often has abnormally low PGE1 values.
- For hangovers and other withdrawal symptoms after excessive alcohol consumption.
- Alcoholic liver damage.
- Chest pains, aches and fluid build-ups (oedema).
- Menstrual pains caused by increased activity of harmful prostaglandins.
- Pre-menstrual tension.
- Hyperactivity in children.

The many beneficial effects of γ-linolenic acid are the object of intense research throughout the world, and several interim results indicate that megadoses of GLA may be able to help in the treatment of some forms of cancer.

My own experiences with GLA have also been positive, particularly in the treatment of atopic eczema, asthma, arthritis and scleroderma. Some allergic children have been helped in the space of only three weeks, while other patients may take two or three months or even longer before they respond to treatment. I always combine GLA with fish oils in order to ensure an adequate balance between these two series of EFAs. Once in a while, there are also patients who do not respond at all to treatment, although this is mostly the case with patients who are undergoing simultaneous treatment with preparations of cortisone and other anti-inflammatory arthritis medicines. These medicines counteract the positive effects of γ-linolenic acid. Nevertheless, it is worthwhile to attempt a course of treatment with the highest recommended doses of GLA over a period of a few months, even though GLA as a food supplement is very expensive. This therapy has become rather more accessible since the arrival on the market of products with a much higher content of GLA per capsule, reducing the number of capsules required to gain the desired effect, and reducing the cost at the same time.

If beneficial effects are gained over a period of time, the high dosage can then be reduced to the minimum dose, which should still be effective.

3. EICOSAPENTANOIC ACID (EPA) AND DOCOSAHEXANOIC ACID (DHA)

What, in the final analysis, are human truths? They are our undisturbed observations.

F. W. Nietzsche (1844–1900)

Two essential fatty acids, EPA and DHA, have been the focus of a great deal of interest in connection with cardiac studies. A few years ago, it was discovered that Eskimos and Japanese fishermen, whose diets are largely composed of fish, exhibit a very low incidence of heart disease. It has now been scientifically investigated, and a statistically significant link between these two factors has been established.

EPA and DHA are important components of human diets, because the organism is not capable of producing them from linoleic acid in sufficient quantities. Good dietary sources of these essential fatty acids are fish and fish oils, linseeds, marine algae and meat from marine mammals (seal and whale meat).

The amount of food which we eat and the way in which we balance our diet is clearly associated with heart diseases. Detailed research has been conducted into the correlation between dietary fat content and heart problems. The risk of hardening of the coronary arteries increases proportionally with the blood levels of LDL-cholesterol (in particular the oxidized o-LDL), while HDL cholesterol has a protective influence.

Cholesterol levels have their effect over the long term, perhaps over decades, while levels of dietary fats have a direct influence on the aggregation of blood platelets and the formation of blood clots. Saturated fats increase the tendency to blood platelet aggregation, while polyunsaturated fats, fish oils in particular, decrease this tendency.

Serious interest in fish oils was first shown after the Danish doctors, **Jørn Dyerberg** and **Hans O. Bang**, compared patterns of illness for Greenland Eskimos and southern Danes. Their studies revealed that diseases such as ischaemic heart disease, psoriasis, asthma, diabetes and disseminated sclerosis were much rarer amongst Greenland Eskimos, while the incidence of cancer was the same for both groups. Conversely, the Eskimos showed a higher incidence of cerebral haemorrhages. The Danish doctors received international acclaim for their work when they discovered that the low incidence of so many diseases in Eskimos was caused by their diet of fish, seal meat and whale meat. This diet is particularly rich in the two polyunsaturated fatty acids, EPA and DHA, which are also known as omega-3-fatty acids.

The effects of EPA and DHA
We know a great deal more today about the way in which EPA and DHA, because of their beneficial influence on the prostaglandin and leukotriene metabolism, prevent and alleviate arteriosclerosis, heart disease, inflammation, high blood pressure, arthritis and migraine headaches.

Prostaglandins and leukotrienes are a large group of local hormones with many different effects in the organism. Prostaglandins, for example, affect the capacity of the blood vessels to expand, while leukotrienes regulate inflammatory reactions. Some prostaglandins promote the formation of blood clots while others inhibit this process. A balance must therefore be achieved between these opposite effects, and it is here that the beneficial influence of fish oils comes into the picture, because they shift the balance away from the tendency to blood clot formation, via their capacity to reduce the aggregation of blood platelets.

EPA and DHA are transformed into prostaglandins in the 3-group, e.g. thromboxane A3 (TXA3) and prostacyclin (PGI3), both of which reduce blood platelet aggregation and blood clotting. EPA and DHA also affect the coagulation capacity of the blood in other ways, for example, they increase the ability of blood corpuscles to pass through the smallest blood vessels. A diet which is rich in fish may therefore be able to prevent ischaemic heart disease. Polyunsaturated fatty acids from the plant world are also beneficial, but do not have the same capacity to prevent blood clots.

At the moment, intensive research is being conducted into the effects of fish oils on other chronic illnesses. A study on the effects of fish oils on psoriasis patients has recently been published in *The Lancet*, and the results were very positive. Other studies have shown benefits from fish oil therapy in atopic eczema, migraines and arthritis. There is now a great deal of evidence to suggest that dietary adjustments in the direction of increased fish consumption is advisable.

The best sources of EPA and DHA are fatty fish, such as herring, mackerel and salmon and it is recommended that fish should be eaten at least twice a week. At the same time, it is advisable to limit dietary intake of animal fats, which directly counteract the beneficial influence of fish oils. If fish is not to one's taste, or if it causes an allergic reaction, then it is possible to take fish oils in capsule form, although it is worth noting the considerable qualitative variations in fish oil products. Fish oils do not elicit allergic reactions, since they do not contain the proteins contained in 'meat' or fish.

At the moment, fish oils containing free fatty acids seem to be the best alternative, because of an excellent bio-availability.

When fish oils are consumed, it is also advisable to ensure an adequate intake of the antioxidants, selenium and vitamin E. Because they are polyunsaturated fatty acids, EPA and DHA are very easily oxidized if the antioxidant defence is not optimal.

LITERATURE

Belch, J. and Ansell D. Effects of evening primrose oil and evening primrose oil and fish oil in rheumatoid arthritis. Brit. Soc. Rheumatol. Business Meeting, 20 November 1986, London.

Bittiner, S., Cartwright, J., Tucker, W. A double blind, randomized placebo-controlled trial of fish oil in psoriasis. *Lancet* **852** (1988) 378–380.

Bjørneboe, A., Søyland, E., Bjørnebo, G. *et al.* Effect of dietary supplementation with eicosapentaeonic acid in the treatment of atopic dermatitis. *Br. J. Dermatol.* **117** (1987) 163–169.

Dyerberg, J. n-3 polyumaettede fede syrer og deres mulige rolle i sygdomspraevention. *Nordisk Medicin*, **103** (1988) 161–165.

Hassam, A. The role of evening primrose oil in nutrition and disease. In: Padley, F., Podmore, J. (eds) *The role of fats in human nutrition.* Ellis Horwood, Chichester, UK, 1985, pp. 84–101.

Horrobin, D. F. *Clinical uses of essential fatty acids.* Eden Press, Montreal-London, 1982.

Horrobin, D. F. The reversibility of cancer: the relevance of cyclic AMP, calcium, essential fatty acids and prostaglandin E1. *Med. hypotheses* **6** (1980) 469–486.

Huang, Y. S., Drummons, R. and Horrobin, D. F. Protective effects of γ-linolenic acid on Aspirin-induced gastric hemorrhage in rats. *Digestion* **36** (1987) 36–41.

Jamal, G., Carmichael, H. and Weir, A. γ-linolenic acid in diabetic neuropathy. *Lancet* **1** (1986) 1098.

van der Merwe, C. γ-linolenic acid: a possible new cure for malignant mesothelioma. Second International Congress on essential fatty acids, prostaglandins and leukotrienes, London, 24–27 March 1985. Abstracts, p. 127.

Salo, M. K. Fatty acids and platelet functions in populations with different rates of ischaemic heart disease. Doctoral dissertation, Series A, Vol. 219, University of Tampere, Finland, 1987.

Ylä-Herttuala, S. Biochemical composition of coronary arteries and aortas in Finnish children and adults. Doctoral dissertation, Series A, Vol. 220, University of Tampere, Finland, 1987.

Glossary

Absorption: the degree to which a substance is assimilated into the blood stream from the digestive system.

Acetaldehyde: *see* aldehydes

Acetyl salicylates: painkilling medicines (aspirin).

Adequate daily intake (ADI): amount of micronutrients, if taken daily, are believed to be adequate for health. See RDA.

Aldehydes: a group of substances and chemical compounds which are related to formaldehyde (formalin). These substances are the by-products of metabolic processes. Aldehydes may cause cross-linkages, mutations and cancer. When fats are oxidized, they produce a toxic substance called *malondialdehyde*, which is also a carcinogen. Most aldehydes are oxidized spontaneously and give rise to the formation of free radicals. The combustion of alcohol in the liver produces acetaldehyde. Formaldehyde (formalin) is also found in tobacco smoke. Formaldehyde is an allergen.

Amino acid: an organic acid which contains an amino group. When several amino acid molecules are joined together in chains, they form polypeptides and proteins. There are 22 known amino acids, of which 8 are essential for human beings. These are: histamine, isoleucine, lysine, methionine, phenylalanine, threonine, tryptophan and valine. They can not be produced by the organism and must be consumed in food.

Amyloid: a type of age pigment which accumulates in the nerve cells. Amyloid is believed to inhibit metabolic processes in the nervous system. In the disease, amyloidosis, this substance accumulates in many tissues beginning in the basal membranes of the hair follicles.

Angina pectoris: heart cramps, chest pains. Mostly arises as a consequence of hardening of the coronary arteries but can have other origins. Can usually be alleviated with preparations of nitroglycerine.

Anorexia: loss of appetite. Anorexia nervosa is obsessive loss of desire for eating.

Antagonist: a substance which counteracts the effects or absorption of another substance.

Antibody: large Y-shaped molecule, produced by the B-lymphocytes in the immune system. It seeks antigens and fastens itself to them. Some antibodies bind themselves to viruses and render them harmless, while others make bacteria and cancer cells more receptive to the destruction of other factors in the immune system.

Antigen: protein which antibodies and other cells in the immune system identify and neutralize. Typical antigens are viruses, bacteria, cancer cells and some protein molecules which can give rise to allergic responses, e.g. pollen.

Antihistamine: *see* histamine.

Antioxidants: substances or chemical compounds which prevent oxidation (also known as rancidification). They attach themselves chemically to oxygen free radicals and substances which produce free radicals. Antioxidants protect the cells by reacting with oxidizing factors and neutralizing them. Conversely, oxidation can damage polyunsaturated fats, proteins and the DNA and RNA in cells. Selenium, zinc, β-carotene, vitamins A, E, C, ubiquinone and some of the B vitamins are natural dietary antioxidants. Many vitamin and mineral therapies are based on the antioxidant effect.

Apoenzyme: the protein component of enzymes, in the form of a coenzyme or as a group of smaller molecules which are attached to an enzyme.

Ascorbic acid: the chemical term for vitamin C.

Arteriosclerosis: hardening of the arteries; *see* plaque.

Atopic: allergic, e.g. atopic eczema is allergic eczema.

Autoimmune disease: arthritis and disseminated sclerosis are examples of autoimmune diseases, where the immune system malfunctions. The risk of contracting these diseases increases with age.

Autoxidation: oxidation which takes place spontaneously and which leads to the rancidification of the cells (lipid peroxidation) and foodstuffs.

Avidin: protein in egg white, which binds itself to biotin and renders it ineffective.

B-lymphocytes: white blood corpuscles which produce antibodies, usually according to directives from the T-lymphocytes. B-lymphocytes are produced in the bone marrow.

Bioflavonoids: also known as vitamin P. These substances are found in fruit peel and in the pith of citrus fruits. They are believed to strengthen the walls of the smallest blood vessels and to improve the function of vitamin C. Rosehips, berries and herbs contain high concentrations of these vitamins.

Biological clock: data which the DNA in the cells has inherited and which determine the life-span of the cells. The biological clock presumably determines the process of ageing, the time of the menopause and balding.

Calcipherol: synthetically produced vitamin D.

Carcinogen: substances which stimulate the initiation phase of cancer.

Carotenoid: carotene, an antioxidant, which occurs naturally in plants. Protects against ultraviolet light, protects experimental animals from cancer and presumably also human beings. β-carotene is found in carrots and gives them their characteristic colour.

Catalase: enzyme which breaks down hydrogen peroxide. It is an antioxidant and is found in all tissues which use oxygen.

Catalyst: a substance which starts off a metabolic process, without either being involved in the process or being changed as a result of the process.

Celiac disease: gluten allergy. Oversensitivity to proteins, especially gluten and gliadin.

Chelate: a loose compound formed between a metal and an organic compound, e.g. an amino acid (amino chelate). Chelation changes the properties of metals and usually improves their bio-availability.

Claudatio intermittens: 'window watchers disease', caused by hardening of the blood vessels in the legs, which prevents the patient from walking further than very short distances at a time.

Coenzyme: the component of an enzyme which is not a protein, usually one of the B vitamins.

Collagen: connective tissue protein, constitutes 30% of all the protein in the body.

Controlled study: an investigative method in which a patient group is compared, for example, to a control group which is not given the treatment under study. In uncontrolled studies it is almost impossible to determine any treatment effect, because it is not possible to ascertain what would have happened had the patient not been given treatment.

Degeneration: impaired function or commencing breakdown of an organ.

Dehydroascorbic acid: a toxic, oxidized form of ascorbic acid or vitamin C.

Desirable plasma levels of antioxidants: concentrations which are suggested ideal for prevention of cancer and cardiovascular diseases.

Diuretic: a substance or medicine which stimulates increased urine secretion.

DNA: deoxyribonucleic acid, which holds the genetic information in the cells.

Docosahexanoic acid (DHA): a polyunsaturated fatty acid which inhibits the aggregation of blood platelets and helps to prevent blood clots. It is found in fish and fish oils.

Dopamine: an important mediator of nerve impulses. Nerve impulses contain information which is passed on by dopamine. This substance is of significance for movement, motivation, instincts, and for the immune system.

ECG (electrocardiogram): a measurement of the activity of the heart.

EEG (electroencephalogram): measurement of electrical impulses in the brain.

Eicosapentanoic acid (EPA): a polyunsaturated fatty acid which inhibits the aggregation of blood platelets and helps to prevent blood clots. It is found in fish and fish oils.

Enzyme: large protein molecule which precipitates metabolic processes. Enzymes consist of two components: the apoenzyme and the coenzyme. The enzyme itself remains unchanged during the metabolic process.

Essential: necessary for life. It is mainly used to describe amino acids, vitamins, minerals and fatty acids which human beings cannot do without in the metabolism and which cannot be produced in the organism, must be taken in via diet.

Gamma-linolenic acid (GLA) (γ-linolenic acid): essential fatty acid which is produced in the healthy human organism from linoleic acid. Production is often limited by various factors.

Glutamate: an amino acid which stimulates the central nervous system. It is often used in Chinese food and as an additive to many other foodstuffs under the name monosodium glutamate, the 'third spice'. It can be dangerous in large quantities, and may cause a reaction known as Chinese Restaurant Syndrome.

Glutathione peroxidase: selenium enzyme which prevents oxidation and destroys free radicals. It is one of the most important antioxidant enzymes in the organism and is dependent on supplies of selenium for optimal activity.

HDL-cholesterol: high-density-lipoprotein cholesterol, the beneficial cholesterol which protects against heart disease and arteriosclerosis.

Hesperidin: a vitamin from the bioflavonoid group.

Histamine: vital for growth and the healing of wounds. Expands the blood vessels and increases the secretion of stomach acids. Large quantities of histamine are harmful and are released in asthma and other allergic responses in the respiratory passages, the skin and other tissues. Antihistamine counteracts this effect.

Homeostasis: condition of balance between the organism and the environment. For example, the organism regulates the absorption of many minerals homeostatically, i.e. it strives to maintain a balance between intake and absorption.

Hormone: a chemical messenger, which has its effect far from the place of production. Hormones are transported with the blood to the place in the body where they have their effects. Examples of hormones are insulin, testosterone, oestrogen and cortisone. Vitamin D has a hormone-like effect.

Hydroxyl radical: a very strong, reactive, cell-destroying radical, formed from the reaction of a superoxide radical with hydrogen peroxide.

Immune system: the organism's defence against infection, other diseases and ageing. It protects the body against arteriosclerosis and cancer by identifying and destroying cancer cells.

Infarction: destruction of tissue because of oxygen deficiency (ischaemia), e.g. heart infarction.

Initiation: the first stage in the development of cancer.

LDL-cholesterol: low-density-lipoprotein cholesterol, which increases the risk of arteriosclerosis if oxidized to o-LDL.

Leukotrienes: end-product of arachidonate metabolism. Leukotrienes occupy a key role in inflammation-like and allergic reactions. They cause a contraction of the respiratory passages which is about 1000 times stronger than that of histamine. They also exacerbate swellings and fluid accumulations.

Linoleic acid: an essential fatty acid which is found mostly in plants, fish and fish oils. It is a component of lecithin.

Lipid peroxidation: rancidification (oxidation) of fats, in particular of polyunsaturated fats. Caused by free radicals and inflammation.

Lipids: fats and oils.

Lipofuscin: brown age pigment which arises from the influence of oxygen free radicals and promotes ageing. It accumulates in the brain and the skin (hence the brown marks which appear on the skin with age).

Lymphocytes: white blood corpuscles which are formed in the lymph nodes, the spleen and the bone marrow.

Lysosome: a component part of the cells which contains harmful enzymes. If the lysosome is damaged and the enzymes are released, it gives rise to tissue damage, e.g. rheumatoid arthritis.

Macrophages: white blood cells which fight viral bacterial and cancer cells. Their function is improved by certain vitamins and minerals.

Malonaldehyde: *see* Aldehydes.

Megavitamin therapy: treatment with doses of vitamins at many times the (RDA) (*mega* means 'huge').

Membrane: thin layer, which consists of fats and which surrounds the cells and lines other organs.

Metabolism: chemical processes in the organism, the breakdown of the components of foodstuffs into a form in which the organism can utilize them.

Microgram (μg): one millionth of a gram.

Milligram (mg): one thousandth of a gram.

Mitochondria: a miniature power station within the cell, which oxidizes nutrients and transforms them into carbon dioxide, water and energy. The mitochondria make use of oxygen free radicals, which are held in check by a protective system. The free radicals become dangerous if this defence system is impaired in its function.

Mutagens: chemical substances which damage the structure of DNA. This alters the functioning of the cell, which can develop into a cancer cell.

Nanogram (ng): one thousand millionth of a gram (one thousandth of a microgram).

Nitrosamines: carcinogens which arise from the reaction of amines with nitrites. They are produced in the intestines and their production can be reduced by the presence of vitamin C.

Nocebo effect: adverse side-effect due to a negative psychological reaction of the patient to a pill or injection (contrast to placebo effect).

Oestrogen: female sex hormone.

o-LDL: oxidized LDL-cholesterol, which promotes arteriosclerosis.

Osteoporosis: brittle bones, particularly affecting women after the menopause. There is no existing treatment but it can be prevented to a certain extent.

Oxidant: oxidizing substance, opposite of an antioxidant.

Oxygen free radical: toxic compound of oxygen with an unpaired electron. Damages or destroys the cell membranes, the genes and the mitochondria.

PABA: para-aminobenzoic acid.

Pellagra: disease caused by deficiency of niacin.

Pernicious anaemia: form of anaemia caused by vitamin B12 deficiency.

Peroxide: oxidized compound which produces free radicals. Starts a chain reaction which leads to the destruction of the cell if adequate antioxidant protection is lacking.

Placebo: 'pretend' medicine, with no pharmacological effect. It is usually given to the control group in order to check whether or not the medicine given to the treatment group has a real effect — or is just a byproduct of the faith which can move mountains. Placebo can sometimes be a good 'medicine' because it works on the psychological plane.

Plaque: hard substance which accumulates on the inside of the walls of blood vessels, contains cholesterol, fats, blood platelets, and blood cells. Plaque formation leads to constriction of the blood vessels.

Progeria: hereditary child disease where the body ages rapidly. Believed to be caused by free radicals.

Promotion: a phase in the process whereby cells begin to behave as cancer cells. Promotion is caused by promoters, chemical substances which affect the DNA in the cells. It may be that the cancer-protective effect of antioxidants comes from the inhibition of the promotion phase.

Pro-oxidant: a substance which promotes oxidation.

Prophylaxis: the prevention of disease.

Prostaglandins: hormone-like fatty acids which occur locally in the tissues and have their effect locally. There are both harmful and beneficial prostaglandins.

Prostacyclin: prostaglandin PGI2. Occurs in the artery walls and inhibits the aggregation of blood platelets and formation of blood clots. Production of this prostaglandin is reduced by peroxidative fats.

Protein: complex molecules composed of combinations of amino acids.

Prudent daily intake (PDI): Daily amount of micronutrients which result in desirable plasma levels.

RDA (Recommended Dietary Allowance): the daily intake of a protein, vitamin or mineral which is required for the maintenance of good health. (Usually based on American norms.)

Reperfusion: resumption of blood supply.

Reperfusion injury: damage to the cells caused by oxygen free radicals. During reperfusion, free radicals react with the cells and form lipid peroxides, leading to cell damage. Antioxidants protect against reperfusion injury.

RNA: ribonucleic acid, which transports information from the DNA in the cell to polyribosomes in the cell nuclei, which then produce protein molecules by copying the information transported from the DNA.

Rutin: an antioxidative vitamin-like substance from the bioflavonoid group.

Serotonin: a transmitter substance in the brain which is responsible for the process of falling asleep.

Superoxide dismutase (SOD): an enzyme which contains copper and zinc (CuZn-SOD) or manganese (MnSOD) and which functions as an antioxidant.

Superoxide radical: oxygen free radical which causes cell changes.

Syndrome: a group of different symptoms which appear simultaneously.

Synergists: substances which function together and reinforce each other's effects, the opposite of antagonists. Many micronutrients function synergistically when taken together.

Synthetic: artificially produced.

Thymulin: hormone produced in the thymus. The activity of thymulin decreases in association with zinc deficiency; the immune system of the elderly is often impaired by decreased thymulin owing to common zinc deficiencies amongst the elderly.

Thymus: a gland which is located behind the breast bone and which programs the T-lymphocytes for the immune system. This gland begins to atrophy after puberty and leads to a gradual weakening of the immune system.

T-lymphocytes: white blood corpuscles which direct the activity of the B-lymphocytes.

Tocotrienols: Vitamin E-like substances, present in the food, for instance in barley and palm oil.

Uric acid: urate, formed in the organism from caffeine and a number of other substances. Accumulations of uric acid in the joints lead to gout.

Vitamins: organic compounds which the organism requires for metabolic processes.

Appendix:
Vitamins and minerals in a range of foodstuffs and recommended dietary allowances

1. CONTENT OF FAT-SOLUBLE VITAMINS (per 100 kilojoules)

The **average** kJ-intake is 8000–10000 for adult women, and 10000–12000 for adult men. The figures are usually higher for people doing heavy physical work and lower for those with sedentary occupations. They are also significantly lower for the elderly.

(1 kcal=4.2 kJ,   1 kJ=0.24 kcal,   1,000 kJ=240 kcal)

Vitamin A
RDA (Recommended Dietary Allowance)=1000 micrograms

Eggs	374
Cheese	128
Milk	79
Skimmed milk	60
Potatoes	6
Apples	16
Cowberries	43
Cabbage	96
Wholemeal flour	2
Rye flour	2
Beef, fried	6
Pork, fried	0
Herring	64

Vitamin E
RDA=10 milligrams

Eggs	2.5
Cheese	0.6
Milk	0.3
Skimmed milk	0.2
Potatoes	0.3
Apples	0.8
Cowberries	0
Cabbage	1.9
Wholemeal flour	0.7
Rye flour	0.6
Beef, fried	0.3
Pork, fried	0
Herring	0

Vitamin D
RDA=5 micrograms

Eggs	2.7
Cheese	0.1
Milk	0.2
Skimmed milk	0.1
Potatoes	0
Apples	0
Cowberries	0
Cabbage	0
Wholemeal flour	0
Rye flour	0
Beef, fried	0
Pork, fried	0
Herring	21,1

2. CONTENT OF WATER-SOLUBLE VITAMINS AND MINERALS (per 1000 kilojoules)

Vitamin C
RDA=60 milligrams

Eggs	0
Cheese	0
Milk	4
Skimmed milk	5
Potatoes	47
Apples	41
Cowberries	22
Cabbage	480
Wholemeal flour	0
Rye flour	0
Beef, fried	0

Pork, fried	0
Herring	0

Vitamin B1 (thiamin)
RDA=1.5 milligrams

Eggs	0.2
Cheese	0
Milk	0.2
Skimmed milk	0.2
Potatoes	0.4
Apples	0.2
Cowberries	0.2
Cabbage	0.7
Wholemeal flour	0.3
Rye flour	0.3
Beef, fried	0.1
Pork, fried	0.8
Herring	0.1

Vitamin B2 (riboflavin)
RDA=1.7 milligrams

Eggs	0.6
Cheese	0.4
Milk	0.8
Skimmed milk	0.9
Potatoes	0.1
Apples	0.2
Cowberries	0.3
Cabbage	0.5
Wholemeal flour	0.1
Rye flour	0.2
Beef, fried	0.3
Pork, fried	0.2
Herring	0.5

Niacin
RDA=19 milligrams

Eggs	5.8
Cheese	4.8
Milk	3.3
Skimmed milk	3.7
Potatoes	5.3
Apples	0.4
Cowberries	1.3
Cabbage	5.8
Wholemeal flour	4.6

Rye flour	2.0
Beef, fried	17.0
Pork, fried	6.1
Herring	14.6

Pantothenic acid
RDA=6 milligrams

Eggs	2.8
Cheese	0.2
Milk	1.4
Skimmed milk	1.8
Potatoes	0.9
Apples	0.4
Cowberries	0
Cabbage	2.0
Wholemeal flour	0.8
Rye flour	0.6
Beef, fried	1.3
Pork, fried	1.3
Herring	0

Vitamin B6 (pyridoxine)
RDA=2.2 milligrams

Eggs	0.2
Cheese	0.1
Milk	0.2
Skimmed milk	0.2
Potatoes	0.8
Apples	0.2
Cowberries	0.1
Cabbage	1.5
Wholemeal flour	0.4
Rye flour	0.3
Beef, fried	0.5
Pork, fried	0.5
Herring	0.8

Vitamin B12 (cyanocobalamin)
RDA=3 micrograms

Eggs	2.7
Cheese	1.0
Milk	1.3
Skimmed milk	1.4
Potatoes	0
Apples	0
Cowberries	0

Cabbage	0
Wholemeal flour	0
Rye flour	0
Beef, fried	1.8
Pork, fried	3.5
Herring	22.8

Folic acid
RDA=400 micrograms

Eggs	39
Cheese	14
Milk	21
Skimmed milk	23
Potatoes	44
Apples	20
Cowberries	11
Cabbage	250
Wholemeal flour	43
Rye flour	60
Beef, fried	18
Pork, fried	6
Herring	0

Biotin
RDA=150 micrograms

Eggs	39.0
Cheese	1.1
Milk	8.3
Skimmed milk	9.3
Potatoes	0.3
Apples	1.2
Cowberries	0
Cabbage	0.9
Wholemeal flour	5.3
Rye flour	4.6
Beef, fried	0
Pork, fried	3.5
Herring	0

Calcium
RDA=800 milligrams

Eggs	89
Cheese	632
Milk	457
Skimmed milk	535
Potatoes	18

Apples	25
Cowberries	162
Cabbage	403
Wholemeal flour	20
Rye flour	17
Beef, fried	17
Pork, fried	10
Herring	162

Potassium
RDA=1000–6000 milligrams

Eggs	203
Cheese	53
Milk	623
Skimmed milk	720
Potatoes	1565
Apples	527
Cowberries	370
Cabbage	3072
Wholemeal flour	296
Rye flour	223
Beef, fried	609
Pork, fried	328
Herring	616

Magnesium
RDA=400 milligrams

Eggs	20
Cheese	21
Milk	46
Skimmed milk	51
Potatoes	75
Apples	26
Cowberries	38
Cabbage	134
Wholemeal flour	99
Rye flour	43
Beef, fried	48
Pork, fried	13
Herring	50

Phosphorus
RDA=800 milligrams

Eggs	328
Cheese	362
Milk	357

Skimmed milk	400
Potatoes	141
Apples	41
Cowberries	73
Cabbage	394
Wholemeal flour	266
Rye flour	146
Beef, fried	340
Pork, fried	211
Herring	456

Iron
RDA=18 milligrams

Eggs	3.9
Cheese	0.2
Milk	0.2
Skimmed milk	0.2
Potatoes	2.1
Apples	0.9
Cowberries	1.7
Cabbage	11.5
Wholemeal flour	3.9
Rye flour	2.3
Beef, fried	4.3
Pork, fried	0.7
Herring	1.4

Zinc
RDA=15 milligrams

Eggs	2.2
Cheese	3.0
Milk	1.7
Skimmed milk	1.9
Potatoes	1.0
Apples	0.3
Cowberries	0.8
Cabbage	2.0
Wholemeal flour	2.7
Rye flour	1.5
Beef, fried	6.1
Pork, fried	1.6
Herring	3.9

Manganese
RDA=3.8 milligrams

Eggs	0.1

Cheese	0
Milk	0
Skimmed milk	0.1
Potatoes	0.7
Apples	0.4
Cowberries	15.1
Cabbage	2.3
Wholemeal flour	2.7
Rye flour	1.6
Beef, fried	0
Pork, fried	0
Herring	0.1

Copper
RDA=2500 micrograms

Eggs	109
Cheese	61
Milk	37
Skimmed milk	42
Potatoes	285
Apples	170
Cowberries	310
Cabbage	346
Wholemeal flour	388
Rye flour	232
Beef, fried	147
Pork, fried	44
Herring	137

Chromium
RDA=50–200 micrograms

Eggs	0.8
Cheese	1.4
Milk	2.1
Skimmed milk	2.3
Potatoes	1.6
Apples	2.0
Cowberries	8.6
Cabbage	4.8
Wholemeal flour	1.5
Rye flour	1.5
Beef, fried	3.6
Pork, fried	3.5
Herring	4.6

Iodine

RDA = 150 micrograms

Eggs	15.1
Cheese	3.6
Milk	15.4
Skimmed milk	17.2
Potatoes	11.9
Apples	6.5
Cowberries	41.5
Cabbage	49.9
Wholemeal flour	0
Rye flour	0
Beef, fried	0
Pork, fried	0
Herring	11.4

Fluorine

Suggested daily allowance = 1000–4000 micrograms

Eggs	47
Cheese	50
Milk	42
Skimmed milk	42
Potatoes	31
Apples	41
Cowberries	43
Cabbage	96
Wholemeal flour	46
Rye flour	39
Beef, fried	36
Pork, fried	23
Herring	228

Selenium

RDA = 70 micrograms

Eggs	17.6
Cheese	2.8
Milk	1.3
Skimmed milk	1.4
Potatoes	0.6
Apples	0.8
Cowberries	0.9
Cabbage	1.9
Wholemeal flour	0.8
Rye flour	0.8
Beef, fried	1.8
Pork, fried	9.4
Herring	41.0

The studies on vitamin and mineral content of various foods are Finnish, because thorough studies have not actually been carried out in very many other countries. The figures may not fit precisely with those of other countries but can certainly be used as a guideline. The RDAs quoted are American.

Index